25.6.94 Bedford St SP

1.20

COMPLETE HOME INSULATION

BARRY WOOD

David & Charles
Newton Abbot London North Pomfret (Vt)

British Library Cataloguing in Publication Data
Wood, Barry
    Complete home insulation.
    1. Dwellings – Insulation
    1. Title
    693.8′32   TH1715
    ISBN 0-7153-7799-X

Library of Congress Catalog Card Number 79-51096

Typeset by Trade Linotype Ltd, Birmingham
and printed in Great Britain
by Biddles Ltd, Guildford
for David & Charles (Publishers) Limited
Brunel House   Newton Abbot   Devon

Published in the United States of America
by David & Charles Inc
North Pomfret   Vermont 05053   USA

Contents

List of diagrams

4

List of tables

1 Why insulate?

Most houses, new or old, are built using materials which lose heat quickly. To keep the house warm, we must pump heat into it continually and, the warmer the house is, the faster the heat escapes. However, heat *can* be stored for long periods —you only have to think of a vacuum flask or a tea pot with a cosy on it—and if you put a layer of insulation right around your house, you can slow down the rate at which it loses heat.

You may wonder why this has not been done before. There are several reasons. When older houses were built, cheap fuels, such as wood and, later, coal, were too readily available. The concept of insulation was not appreciated and, indeed, really satisfactory materials for insulating houses were not available. Although houses built in the last decade or so could easily have been built in such a way that they lost heat slowly, the Building Regulation standards to which they were built have, in the past, not specified insulation as being of particular importance. They are being gradually improved but, even now, the standards of insulation called for in these regulations are far too low.

If you are having a new house built it is quite possible for your architect to include materials which only let heat through at a slow rate, although unfortunately it is not possible to stop the loss altogether. However, most architects and builders have not yet got around to this way of thinking and it is up to you, the customer who is paying for their services, to insist on a satisfactorily insulated house.

The remedy is simple—keep the heat in and the heaters can

then be turned down or off. The fuel consumption and the consequent bills are then reduced to a minimum.

Central heating systems cost money to install, run and maintain. All heating appliances burn fuel and, considering inflation and the ever increasing cost of fuel, the average householder can expect to pay out over £25,000 on heating bills during the course of his lifetime.

The inclusion of central heating is currently regarded as an asset when it comes to selling a house—whereas it ought to be regarded as a liability! With the ever increasing cost of fuel we must change our thinking and move away from conventional central-heating systems. Why spend money just to burn money?

People tend to think that insulation should cost very little; unfortunately, this is not true. To achieve a good result, you must spend as much on your insulation system as you would have spent on a good central-heating system. However, while insulation costs money to install, it costs *nothing* to run and maintain. The benefits of correct insulation are, simply, a warm house in the winter, a cool house in the summer and low heating costs.

This book aims to provide the necessary information to enable both someone building a new house and someone who wants to insulate their existing one, to do so to high standards of insulation. There is nothing new about the materials and techniques involved, in either case, and all the materials should be readily available.

The vast majority of houses in this country are of conventional brick or stone construction and this is the only type of house considered in detail here. If, however, you are buying a new timber or prefabricated house, then you should make sure that the insulation thicknesses and specifications are up to the standards recommended in chapter 2.

In the construction of a new house, most of the insulation is installed after the basic shell has been completed. The cost of the insulation will be included in the total cost of the house and will be paid for through your mortgage. Councils may be giving grants towards the cost of installing some types of insulation and it could be well worth your while to check with the local council before you start the work.

For existing houses the instructions are given in the order of the jobs which can be completed most easily first. They are fully detailed and you can either do most of the work yourself, as many of us are keen to do nowadays, or have it done professionally, to the standards given. Tackle the job properly —you cannot do it all at once. Work down from the top of the house and, because heat rises, you will find that the upstairs rooms need no heating at all for the average person. The cavity wall insulation can be done immediately but the wall panelling and ground-floor insulation should be done gradually as each room becomes due for redecoration.

If you are buying a new house, about to redecorate, thinking of buying extra heaters, finding that your house is too cold or your fuel bills too high, or thinking of installing central heating, then this book is for you. The days of cheap organic fuels are rapidly coming to an end—North Sea gas is already nearly three times the price of the old manufactured gas and there is a huge increase in the price of oil because of its potential scarcity. The effectiveness of the insulation recommended is such that the need for central heating as it is currently known is eliminated and your fuel consumption for heating purposes will be reduced to a minimum. Don't burn all the fuel now—SAVE IT for future generations!

2 Keeping warm

Most common structural materials allow heat to pass through them relatively easily; stone, for example, feels cold because the heat from your hand passes into it rapidly. The amount of heat which they let through increases as the area increases and as the temperature difference between the inside and outside increases. The loss decreases with the thickness of the material.

By and large, most walls, windows, roofs and floors are built to similar thicknesses and from similar materials. In practice, the difference in heat lost between houses in the same street is simply due to the difference in size. There are, however, advantages to be gained if you are having a new house built or are moving into another one. Choosing one on a south-facing slope will mean that the house is warmed by the sun for more of the day than one on a north-facing slope. It will also help if the house is sheltered from strong winds. Even with a semi-detached house, the half having the more southerly aspect will be warmer than the other half.

Air is a good insulator provided that it is not allowed to circulate and insulation materials contain a large amount of still, trapped air. Materials like fibreglass, mineral fibre, polystyrene or urea foam are just materials which can be conveniently expanded or foamed to contain the air. Insulation materials feel warm to the touch because they allow less heat to pass into them than that which your hand is already losing to the air which normally circulates around it. Good insulation materials in fact contain something in the region of 98 per cent air.

Heat loss figures

To enable comparisons to be made between different exterior surfaces of houses, measurements have been made to assess the rates at which they let heat through. This heat loss figure is measured in watts per hour for every square metre of surface area and for every degree centigrade difference between the internal and external temperatures. Of course, the higher the heat loss figure, the faster the structure loses heat, and the faster heat has to be pumped into the house to maintain the desired temperature and the higher are the fuel bills.

Average heat loss figures are shown below, to enable you to see the difference between various parts of your house. Figures are given for older houses which have no insulation at all, houses built to the Building Regulation (1976) standards and for correctly insulated houses. You will note the huge reduction in values for houses which are properly insulated.

TABLE 1
TYPICAL HEAT LOSS FIGURES

	roof	floors	walls	windows
Older uninsulated houses	1·9	0·7	1·6	5·7
Same houses properly insulated	0·23	0·23	0·3	2·0
Reduction in heat loss	88%	67%	80%	65%
Newer houses, built to Building Regulation (1976) Standard	0·6	1·0	1·0	5·7
Same houses, insulated to higher standards	0·18	0·23	0·21	2·0
Reduction in heat loss	70%	77%	79%	65%

The low heat loss figures are achieved by adding insulation of a suitable thickness to the basic structure of the house to obtain the desired effect. The regulations have also imposed a limit on the amount of heat lost through the whole wall area including the windows. Because heat is lost more rapidly through the windows than through the walls some architects achieve the target simply by restricting the size of the windows—a measure obviously not to be encouraged! If you have checked the previous table carefully you will see an apparent discrepancy, in that the heat loss figure for the floor of an older house is lower than that for a new one. This is because the regulation figure is set fairly high and most common floor constructions meet the standard easily.

Insulation recommendations

The insulation standards set in this book are different for existing and new houses. A higher standard is worked to for new houses because it is much easier to incorporate the correct thickness of insulation whilst your house is still on the architect's drawing board than to add on to an existing house. Houses are built to last for a long time and the insulation standards of a new house must be good for the year 2000 as well as 1980. The following table shows the ideal standards of insulation thicknesses for both new and existing houses. The reasons for selecting these standards are set out in the appropriate sections later in more detail.

TABLE 2
RECOMMENDED INSULATION THICKNESSES

	New houses	Existing houses
	mm	mm
Roof	200	150
External walls	150	100
Ground floors	100	100
Windows	Triple glaze	Triple glaze where possible, otherwise double glaze
Outside doors and opening windows	All closing surfaces to be fitted with draught stripping	

3 Summary of heat losses and costs

Each of the later chapters in this book is devoted to one aspect of the exterior surface of your house and it includes detailed calculations showing the heat losses both before and after insulation. These losses have been summarised in the tables following to show the total rate at which heat is lost every hour from a normal-sized house in winter. From this we can work out the size of the heating system needed to keep the house warm. We can also take this a stage further and work out the power consumption for a year and estimate the fuel bills.

We shall see that a correctly insulated house—even a fairly large one—needs little heat to keep it warm and consequently the running costs of the heating system are very much lower than we have been accustomed to paying in the past. The unit of heat used throughout is the kilowatt hour (kWh). This is, quite simply, the amount of heat given off by a one kilowatt (kW) electric fire in one hour. It is the unit in which your electricity is metered and charged and easy comparisons can be made between the costs of different fuels.

Heat gains

An allowance is also included for the heat which is produced inside the house from sources other than the heating system. These are called heat gains. All electrical appliances turn the electricity which they consume into heat which is dissipated inside the house. Electric lights, for example, produce 98 per cent heat and 2 per cent light from the electricity they burn.

15

Refrigerators and deep freezes take heat from their contents and give it out from the radiator fixed to the back of the cabinet. If you place your hand on this radiator when the refrigerator is running you will be surprised at how warm it is. Television sets and water heaters produce quite a lot of heat. Gas and electric cookers can easily consume and give out up to 6 kW whilst they are in use.

Even people produce significant amounts of heat. Because the human body operates at a temperature well above that of a comfortable room we lose heat to the air from uncovered areas of the body like the hands and face, as well as breathing out heated air. An active person can dissipate about 300 watts and it only needs a few energetic children running around to produce the same effect as a small electric fire!

The figure shown for these gains in heat is an average over a period of time. It increases with the size of the house on the assumption that there will be more occupants in a large house than in a small one.

Total heat loss from a house in winter

Each of chapters 5, 8, 12 and 15 contain details of the heat losses from a particular part of the house in winter and these have been brought together in the next two tables to show the totals. Once the heat loss has been determined you will see that a percentage has been added on to arrive at a suitable size for the heating system. This is partly to give a margin for really cold weather and partly to give the system some extra capacity for heating the house up if it has been left empty for a while, if, for example, you go away for a holiday in the winter. The amount added to the system for the insulated house is larger because the system is smaller. If, instead of insulating the house, you had chosen to install a central-heating system, then the firm engaged to do this would have added on about 5 kW to allow for water heating and they would have installed boilers of about 10 kW, 13 kW, 17 kW, and 21 kW respectively for the four sizes of houses.

TABLE 3
AMOUNT OF HEAT LOST EVERY HOUR IN COLD WEATHER FROM NEWER HOUSES

	Typical House size			
	2-bed-roomed terraced	3-bed-roomed semi	4-bed-roomed detached	5-bed-roomed detached

Houses built to Building Regulation (1976) standard of insultation

Through—	kW	kW	kW	kW
Roof	0·3	0·45	0·7	0·9
Windows	1·0	1·6	2·6	3·2
Draughts	1·0	1·3	2·1	3·0
Ventilation	1·0	1·3	2·1	3·0
External walls	0·9	1·4	2·5	3·0
Ground floor	0·4	0·55	0·9	1·2
TOTAL	4·6	6·6	10·9	14·3
Deduct heat gains	0·6	1·0	1·6	2·0
NET LOSS	4·0	5·6	9·3	12·3

Add 25% OUTPUT OF HEATING SYSTEM

REQUIRED	5 kW	8 kW	12 kW	16 kW

New houses built to high standards of insulation

Through—	kW	kW	kW	kW
Roof	0·1	0·15	0·2	0·3
Windows	0·35	0·6	1·0	1·15
Draughts	Nil	Nil	Nil	Nil
Ventilation	1·0	1·3	2·1	3·0
External walls	0·2	0·3	0·5	0·6
Ground floor	0·15	0·2	0·3	0·45
TOTAL	1·8	2·55	4·1	5·5
Deduct heat gains	0·6	1·0	1·6	2·0
NET LOSS	1·2	1·55	2·5	3·5

Add 50% OUTPUT OF HEATING SYSTEM

REQUIRED	2kW	3 kW	4 kW	6 kW

TABLE 4
AMOUNT OF HEAT LOST EVERY HOUR IN COLD WEATHER FROM OLDER HOUSES

	Typical House Size			
	2-bed-roomed terraced	*3-bed-roomed semi*	*4-bed-roomed detached*	*5-bed-roomed detached*

Older houses with no insulation at all

Through—	*kW*	*kW*	*kW*	*kW*
Roof	1·0	1·3	2·1	3·0
Windows	1·0	1·6	2·6	3·2
Draughts	1·0	1·3	2·1	3·0
Ventilation	1·0	1·3	2·1	3·0
Walls	1·5	2·1	4·0	4·8
Floors	0·4	0·55	0·9	1·2
TOTAL	5·9	8·15	13·8	18·2
Deduct heat gains	0·6	1·0	1·6	2·0
NET LOSS	5·3	7·15	12·2	16·2

Add 25% OUTPUT OF HEATING SYSTEM

REQUIRED	7	10	15	20

Older houses correctly insulated

Through—	*kW*	*kW*	*kW*	*kW*
Roof	0·1	0·15	0·25	0·35
Windows	0·35	0·6	1·0	1·15
Draughts	Nil	Nil	Nil	Nil
Ventilation	1·0	1·3	2·1	3·0
Walls	0·3	0·45	0·8	1·0
Floors	0·15	0·2	0·3	0·45
TOTAL	1·9	2·7	4·45	5·95
Deduct heat gains	0·6	1·0	1·6	2·0
NET LOSS	1·3	1·7	2·85	3·95

Add 50% OUTPUT OF HEATING SYSTEM

REQUIRED	2	3	5	6

You can see from these two tables that the reduction in the size and output of the heating system needed for the insulated house is very substantial. It is indeed such that the installation of central heating, as it is currently known, is unnecessary except, perhaps, in the largest of houses.

Power consumption for a whole year

Using figures from table 3, we can carry our calculations a stage further to work out an estimate of how much power an ordinary house is likely to use in a year, together with the reduced amount for the correctly insulated house.

An ordinary house needs heating for about 230 days a year, mainly during the late autumn, winter, and early spring. In the summer very little heating is needed. An insulated house on the other hand only needs to be heated for about 180 days because the heat from the appliances and the occupants—the heat gains—are quite sufficient to warm the house in mild weather. Most people switch the heating system off overnight and allow the temperature to fall: sixteen hours of heating a day is usually sufficient. In addition, the average outside temperature in winter is above freezing and the heat loss for the year can be reduced by one third to allow for this.

The amount of power used for heating a three-bedroomed semi-detached house for a year, using the figure from table 3 is, therefore:

$$5.6 \times 230 \times 16 \times \tfrac{2}{3} = 13,000 \text{ kW hrs}$$

For the correctly insulated semi-detached house, the figure would be reduced to:

$$1.55 \times 180 \times 16 \times \tfrac{2}{3} = 3,000 \text{ kW hrs}$$

These figures appear in table 5.

We have to remember that we have been talking only about the amount of power used for heating the house so far. To this we must add the power used for running all the electrical and gas appliances and that used for heating water. This we can obtain from the heat gains figure in table 3, because, after all, this was an estimate of the heat given off by these

19

appliances and, for practical purposes, it is the same as the power consumed by them. These appliances are in use all the year round—365 days—and the power consumed is the same for both an insulated and an un-insulated house. For the semi-detached house it is therefore:

$$1 \cdot 0 \times 365 \times 14 = 5,000 \text{ kW hrs}$$

The appliances will not be in use for quite so long as the heating system because various members of the family are out of the house from time to time and a slightly reduced figure of fourteen hours is used for this calculation. All these figures appear in table 5 and you can easily check the rest for yourself if you so wish.

TABLE 5
ESTIMATE OF YEARLY POWER CONSUMPTION

	2-bedroomed terraced	3-bedroomed semi	4-bedroomed detached	5-bedroomed detached
Typical house size				

Houses built to 1976 Standard of insulation

kW hrs

For heating	9000	13000	22000	30000
For lighting etc	3000	5000	8000	10000

Correctly insulated houses

kW hrs

For heating	2500	3000	5000	7000
For lighting etc	3000	5000	8000	10000

Again by comparing the two sets of figures you can clearly see the huge reduction in power consumption of the correctly insulated house. In fact, the amount of power used for heating has been reduced below that used for lighting and other purposes.

Running costs for a year

We are now nearly in a position where we can see just how much the power consumption figures in table 5 represent in terms of money. Previously we said that all the power figures were to be in kilowatt hours and so we must convert all the different ways the fuel companies have of charging for their fuels into a price per kilowatt hour to enable us to make comparisons. Electricity is the easiest because it is already charged in this way. We must also take into account how usefully we use the fuel; for example, on-peak electricity is the most efficient as practically all the power coming into the house is usefully used. However some of the heat escapes from the house overnight and it is probably about 90 per cent efficient. The same ought to apply to off-peak electricity but in this case the supply is only available over-night and some of the stored heat is wasted. It is probably 80 per cent efficient.

Oil, solid fuel and gas boilers all waste heat in the flue gases which go up the chimney and they are at most 70 per cent efficient. Gas fires range from about 40 to 60 per cent and an open solid fuel fire can give out as little as 20 per cent of the potential amount of heat in the fuel as useful heat to the room.

The following conversion figures take the differing heat content of the fuels and an average efficiency with which we use them into account.

TABLE 6
CONVERSION FIGURES FOR FUEL COSTS

Fuel	Pricing	Conversion figure
Electricity		
on-peak	pence per unit	1·1
off-peak	pence per unit	1·2
Oil	pence per litre	0·15
North Sea gas	pence per therm	0·06
Solid fuels	pounds per 50 kg	0·5

The price of fuel varies from area to area but at probable 1980 prices the following approximate comparisons can be made.

TABLE 7
COMPARISON OF FUEL COSTS

Fuel	Price	Conversion figure	Cost per kW hour
Electricity			
on-peak	3·3p per unit	×1·1	3·8p
off-peak	1·3p per unit	×1·2	1·6p
Oil	15p per litre	×0·15	2·3p
North Sea gas	22p per therm	×0·06	1·4p
Solid fuel	£3.30 per 50 kg	×0·5	1·7p

As an average figure, the cost of the power used for lighting, cooking, heating water and so on will be taken as being two thirds of the cost per unit of on-peak electricity. Lighting, for example, nearly always uses electricity at peak rates, the heat from our bodies is free and the cost of water heating and cooking varies from house to house depending on the fuel used. At probable 1980 prices the figure used in the next table is therefore

$$3.8p \times \tfrac{2}{3} = 2.5p \text{ per unit}$$

You should work out the cost of the fuels you are using at up-to-date prices for your area to make true comparisons, but the table opposite has been worked out, for interest, using the figures above.

Looking at table 8 you see that substantial reductions in the running cost of your home can be made by insulating it correctly. The saving in cost on the power used for heating alone can be up to 80 per cent. The overall saving, taking into account the power used for other purposes as well, is from 45 per cent to 65 per cent depending on the combination of fuels used.

TABLE 8
ESTIMATE OF YEARLY POWER COSTS—
1980 PRICES

	Typical house size			
	2-bed-roomed terraced	3-bed-roomed semi	4-bed-roomed detached	5-bed-roomed detached

Houses built to 1976 Standards of insulation

Cost of the fuel for heating only using—				
Electricity	£	£	£	£
on-peak	340	500	820	1100
off-peak	140	210	350	480
Oil	200	300	500	700
North Sea gas	120	180	310	410
Solid fuel	150	220	370	500
Add on the cost of fuel used for light etc at 2·5p	70	120	200	250

Correctly insulated houses

Cost of the fuel for heating only using—				
Electricity	£	£	£	£
on-peak	100	120	200	270
off-peak	40	60	80	120
Oil	60	70	120	170
North Sea gas	40	50	70	100
Solid fuel	50	60	80	120
Add on the cost of fuel used for light etc at 2·5p	70	120	200	250

At the moment, most of your money spent on fuel bills is for heating. This will be reduced to such an extent that most of the money spent on future bills will be for auxiliary purposes such as lighting, heating water and cooking. We are all the time being persuaded to purchase colour television sets, deep freezes and powered appliances and nearly all of these consume electricity at peak rates. As you will now have made substantial savings on your heating bills it will be up to you to use your house appliances to the best effect with a view to reducing their running costs. It is quite possible that manufacturers will have to develop domestic appliances in the coming years so that they use less power.

For interest, you can compare these tables with your present fuel bills for the past year. Do not forget to add together all your bills for electricity and gas, coal or oil. The total for your present house will not be exactly the same as the figures in the table because the latter are estimates. People use their houses in different ways and a bachelor for example would very likely use far less power than would a family living in the same size of house.

4 Ventilation

The remedy for a stuffy room is simple—open a window slightly. In a properly draught-proofed house this will only ventilate the room concerned and will not cause a gale to blow through the rest of the house (unless of course you have left the back door open). Because the house is so uniformly warm you will find that a window can be opened even on cooler days when, at the moment, you would be keeping them all closed. The internal doors can always be left slightly ajar to allow air to circulate around the house.

It is not possible to seal completely all the gaps around the doors and windows and a small amount of ventilation will take place in any case. Windows should be opened every morning to clear stale air and it is important that this is done as it also allows any build up of moisture in the house to be dispersed. Bathrooms create a lot of steam and the window should be opened for a short while after use. Running an inch or so of cold water into the bath before turning on the hot tap will work wonders in reducing the amount of steam. In a new house make use of the night ventilators fitted into the window frames. Kitchens can also produce a lot of steam and this can be cleared preferably by just opening the window slightly; a more expensive way of doing it would be to install an extractor fan.

A much better approach is to create less steam in the first place, and this can be achieved in several ways. A pressure-cooker could be used in place of open pans (it uses less fuel as well). Once pans start to boil they can usually be turned down to a gentle simmer. Food does *not* cook more quickly at fast boil—the temperature of the water remains at exactly 100°C; it only wastes heat and creates clouds of steam. Make sure pans have lids on to keep the steam in.

A modern automatic washer does not produce a lot of steam but you could try it at a lower temperature setting. This will save fuel as well as speeding up the cycle. If you are buying a new washer make sure that it has a fast spin speed—at least 800rpm. Some will do 1000rpm and this extracts more moisture from the clothes. When washing with a twin-tub machine, or by hand, open the window a little to let the steam escape. Tumbler driers should always be installed correctly with a vent pipe to take the steam outside.

5 Heat losses through the roof

Most houses have a pitched roof covered with either man-made tiles or natural slates. The slope allows water to run off quickly and gives the roof good weather-resisting properties. In newer houses there is a layer of felt under the tiles, the purpose of which is to reduce the amount of wind blowing through the loft space and to divert any water which penetrates under the tiles, should they be damaged. Older houses do not have this felt and the slates are sealed from beneath with mortar. This tends to drop off as the house ages, allowing quite strong draughts to blow through the loft.

In either case the materials are thin and allow heat rising from the ceiling beneath to pass through easily. The ceiling will be either plasterboards or lath and plaster and again heat rising from the room below passes through these materials easily. To prevent this loss of heat a layer of insulating material must be installed in the loft. The insulation will be thick, compared with the roofing materials, and its exact location will depend on whether or not the loft is boarded for use as an attic or dormer room.

Once the insulation is installed, the loft in the winter will become quite cold and the conditions in the loft can be compared with those underneath a timber-suspended floor. Provision is made for underfloor ventilation to remove dampness, by building airbricks into the walls. Similarly, with the roof, it is absolutely essential to ensure that there is adequate ventilation in the loft and this is dealt with in chapter 6.

Heat loss figures

The heat loss figure decreases as the thickness of insulation increases and the following table shows typical values for both insulated and un-insulated roofs. The figure for the older roof is for one having felt under the tiles. For an older slate roof, sealed with mortar, the loss is about 30 per cent higher.

TABLE 9
ROOF HEAT LOSS FIGURES AND INSULATION THICKNESSES

	Insulation thickness mm	Heat loss figure	Reduction in heat lost— newer roof %	Reduction in heat loss— older roof %
Older roof	—	0.0019	—	—
Roof to 1976 standards	—	0.0006	—	—
Same roof with insulation	50	0·00055	10	70
	75	0·00041	31	78
	100	0·00033	45	82
	150	0·00023	60	88
	200	0·00018	70	—

For a new house a very high standard of insulation should be aimed for and the thickness of the insulation is to be a minimum of 200mm. Current roofs meet the 1976 standard with less than 50mm of insulation and the correctly insulated roof loses 70 per cent less heat than these. For an existing house an insulation thickness of 150mm is recommended. A slightly lower standard is worked to, because other parts of the house cannot be insulated as well as a new house, at reasonable cost. Any small savings made by increasing the roof insulation thickness would not be matched by similar small savings from other parts of the house to build up into a worthwhile amount.

Heat loss calculations

To work out the amount of heat lost through the roof we need to multiply together the heat loss figure, the ceiling area and the temperature difference between inside and outside the house. The heat loss figure can be obtained from the previous table and you will see that all the values have been divided by one thousand compared with those in table 1. This is so that the answer is in kilowatts which is a sensible, easily understood value rather than in watts which is too small a value for

TABLE 10
AMOUNT OF HEAT LOST THROUGH THE ROOF

	Typical House Size			
	2-bed-roomed terraced	3-bed-roomed semi	4-bed-roomed detached	5-bed-roomed detached
Ceiling area in square metres	30	40	65	90
Older un-insulated roof Heat loss				
Per hour kW	1·0	1·3	2·1	3·0 ($\times$ 0·0019 $\times$ 18)
Per day kW hrs	16	20	33	48 ($\times$ 16)
Partly insulated roof to 1976 standards Heat loss				
Per hour kW	0·3	0·45	0·7	0·9 ($\times$ 0·0006 $\times$ 18)
Per day kW hrs	5	7	11	14 ($\times$ 16)
Older roof with 150 mm of insulation Heat loss				
Per hour kW	0·1	0·15	0·25	0·35 ($\times$ 0·00023 $\times$ 18)
Per day kW hrs	1·6	2·4	4	6 ($\times$ 16)
New house with 200 mm of insulation Heat loss				
Per hour kW	0·1	0·15	0·2	0·3 ($\times$ 0·00018 $\times$ 18)
Per day kW hrs	1·5	2	3·5	5 ($\times$ 16)

practical purposes. The ceiling area is easily measured and for comfort the upstairs rooms should be at 18°C. The temperature outside when it is freezing is 0°C and the temperature difference is therefore 18°C. Because most people switch the heaters off overnight, a house needs heating for about sixteen hours a day.

Table 10 summarises the losses through the roof for various typically sized houses and also shows how the calculations have been worked out. Unless your house is unusually large, one of the sets of figures will roughly correspond to your house. If you so wished, you could measure up your house and work out a set of figures for it.

Looking down table 10 for your particular size of house, you will be able to see how the amount of heat lost through your roof decreases as the thickness of the insulation increases, and the extent to which this is so. If you wanted to see what this represented in cost terms, you could do some calculations similar to those in chapter 3.

It is estimated that a third of British houses still have completely un-insulated roofs. Is yours one of these? If so, you will see, in the following pages, how easy it is to insulate a simple roof.

6 Insulating an unfloored loft

This is the most usual and simplest roof construction to insulate. The insulation is laid in two layers—the first between the joists and the second over the joists, to give the required thickness. The method is similar for both new and existing houses. At the same time, the cold water tank and any pipework, if they are in the loft, must be insulated to prevent them from freezing in the winter because the loft space will be very cold once insulated as there is little heat rising into it.

The warm air inside a house tends to absorb more moisture in the form of water vapour than the cooler air outside can hold. Most ceiling materials are fairly porous and allow water vapour to pass through them. To prevent water vapour passing through the ceiling of a new house into the cold roof space, where condensation might take place on the cold roof structure, a layer of material impervious to water will be included in the construction.

A certain amount of ventilation is also required in the loft. It is not possible to do this with existing houses but older roofs tend to have gaps in them which allows wind to blow through the loft space dissipating the moist air. It is absolutely essential that this flow of air is maintained, or even increased, and is not restricted in any way.

Ceiling and loft construction for a new house
The ceilings will, of course, be installed by the builder to your architect's instructions. To provide a substantial vapour barrier a layer of polythene sheet, 125 microns thick, should be tacked to the underside of the ceiling joists. So far as possible, this

sheet should be in one piece; it should overlap the walls slightly and any joints are to be sealed with polythene tape. Foil-backed plasterboard should be used for the ceiling and the excess polythene trimmed off after the plaster boards have been fixed. Naturally this only applies to the ceilings directly beneath the loft; ceilings between ground-floor rooms and first-floor rooms do not need a vapour barrier.

New houses have a layer of felt under the tiles and this effectively prevents any flow of air through the loft. Ventilation bricks must be built in, either under the eaves, or in the end walls—if it is a detached house—to ensure adequate ventilation of the loft.

MATERIALS REQUIRED

There are two sorts of insulation materials suitable for laying between the joists. In the first place, there are fibreglass or mineral fibre rolls 400mm wide. These are to be laid to the full depth of the joists, and their thickness must be chosen accordingly. If the joists are deep, two layers may be needed. Secondly, you can use vermiculite granules which are poured in, to the full depth of the joists.

For laying over the joists, either fibreglass, or mineral fibre rolls 1200mm wide, are suitable. The thickness should be chosen to bring the total thickness of insulation up to a minimum of 200mm for new houses or 150mm for your present house.

For insulating the water tank a proprietary kit made from fibre-filled quilt or polystyrene sheet is used. Pipe insulation is made either in the form of a glass or mineral fibre roll which is wrapped round the pipes, or as flexible foamed or rigid-fibre tubes, which are split so that they can be fitted over the pipes. These tubes have different bore sizes and it is important that you use the correct size. The two most popular sizes are 15mm and 22mm; this refers to the outside diameter of the copper pipes they are designed to fit. If the pipes are lead you should use the fibre pipe-wrapping material.

For a new house, your architect will specify all the materials, work out the quantities needed and arrange for their purchase and installation.

For an existing house, the ceiling area can be measured quite easily, and the number of rolls or bags of insulation required, worked out. This varies from maker to maker, depending on how the insulation is packed, but the area which the material will cover is shown either on the bag or in a leaflet which your supplier will give you.

The tank insulation kits are also made in various sizes and you should measure the tank before purchasing the kit. The kit might have to be larger than the tank to accomodate any pipework adjacent to the tank. For pipe insulation, measure the length of exposed pipework in the loft and round up the figure to allow for bends and joints.

All these materials can be purchased from plumbers or builders' merchants, most Do-it-Yourself stores and some supermarkets.

Insulating the roof of a new house

Your builder will be doing this work to your architect's instructions and the following is a brief summary of the procedure required from them.

The first layer of insulation is to be laid snugly between the joists and must be level with, or slightly proud of the top of the joists, to allow for settlement. The material must stop short of the eaves so as not to block the loft-ventilating bricks. No insulation should be laid under the cold-water tank, if it is in the loft, as a small amount of heat must be allowed to rise into it to prevent it from freezing in the winter.

The second layer of insulation, 1200mm wide, is to be laid across the run of the joists until the whole floor area is covered. Runs of the material are to be butted snugly together and the material is to be cut to fit snugly against any brickwork. Again, care must be taken that this second layer does not block the gaps between the rafters and so cut off the loft ventilation. A section through the completed insulation is shown in diagram 4.

The cold water tank and any pipework in the loft are also to be insulated.

Insulating your existing house

The insulation is, in fact, exactly the same as for a new house but, because you will very likely be doing the work yourself, it is considered in much more detail. Like all things, there are right and wrong ways of doing jobs and precautions to be observed.

PREPARATIONS AND PRECAUTIONS

In an unfloored loft you must step only on the wooden joists. The plaster between the joists comprises the ceiling of the room below and it is quite thin. If you tread or kneel on it you will go straight through and have an expensive repair to make. For the same reason tread lightly on the joists and do not stamp about or the ceiling plaster may crack. It is better to take up a few strong boards to stand on: span these across the joists, but be careful not to stand on an end which may tip up. Do not rush, and mind your head on low beams.

Take up a car-type inspection lamp and hang it on a hook screwed into a beam as high as you can reach. A good light is essential and a torch is unsatisfactory as it is all too easily knocked over when you are working in a confined space.

Fibreglass particles will prick the hands and, whilst it is not harmful, the effect can be very irritating for a while, so wear a pair of gardening gloves and make sure your wrists are well covered. Lastly, you should not go into the loft if you are likely to be taken ill by reason of age or other cause—let someone else do the work.

CHECKS TO MAKE IN THE LOFT

Look around for wet or rotting timbers; all the beams should be firm and dry. If any are wet or if you can push a medium screwdriver into any of them without difficulty, especially near the ends, seek professional advice. Similarly, if you find lots of little round holes in the timberwork, and the holes look clean, again, seek advice, as they will have been made by woodworm.

Electric cables covered in a light grey plastic should be all right as they will be the modern type. If the cables are covered with black rubber or black or red waxed cotton material, have them checked by an electrician because they are obsolete and need replacing. Do not stand on cables or junction boxes as you may well damage them without realising it.

Check to see if there are any obvious gaps in the slates or tiles, then check to see if the loft is adequately ventilated. If the house has a slate roof, sealed with mortar, there will, more than likely, be a quite adequate draught through the loft. A good idea is to go up on a windy day and to check that you can feel draughts blowing in under the eaves and between the slates at intervals.

Newer houses have felt under the tiles and this effectively blocks off any air flow through the loft. Again, go up on a windy day and feel to see whether there is wind blowing in from under the eaves. If there is little or no draught then it *must* be increased by chipping out an inch or so of cement from between the top layer of bricks under the eaves as shown in diagram 1. Do this between one pair of bricks, say between every third pair of rafters. An alternative method would be to knock out a brick every 2–3m and replace it with an air brick. This should be done on both sides of the house. A further alternative is to put two air bricks in each end gable wall, if it is a detached house.

COLD WATER TANK AND PIPES

Fit a new washer to the ball valve and check that it is working correctly. Do the same for the central heating header tank if it is also in the loft. Both of these tanks are going to be covered in and there is little point in having to strip the insulation off in a few months time just to replace a leaky washer.

Fit the pipe insulation along the pipes, cutting it at the bends to make a neat fit. Fit the proprietary kit around the cold-water tank or tanks, following the instructions supplied with the kit. As well as preventing the tank from freezing it will also keep out dust and dirt.

LAYING THE INSULATION

For the first layer, using glass or mineral fibre rolls 400mm wide, partly unroll the roll along the top of two joists, starting from the eave as shown in diagram 1. Tuck the insulation between the joists and under the eave *only* until it is level with the bottom side of the rafters. The spaces between the rafters must be clear right down to the eave so as not to restrict the ventilation. Roll the material out along the joists tucking it in lightly as you go, as in diagrams 2 and 3

Continue over the whole area, filling in all the spaces between the joists and the brickwork as well. Joins between the rolls should be butted snugly together. Try to lay the material in

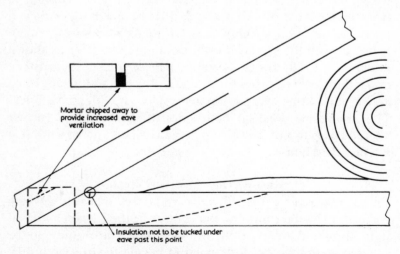

Diag. 1 Rolling out the insulation—new or existing house

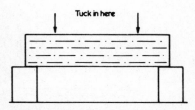

Tuck in here

Diag. 2 Tucking in the insulation—new or existing house

Diag. 3 Insulation when tucked in—new or existing house

continuous lengths by easing it under things like cables, otherwise cut it with sharp kitchen scissors and butt the two ends together under or at the obstruction. Do not lay insulation under the cold water tank as a small amount of heat must be allowed to rise into it.

Do not flap the insulation about otherwise a lot of dust will be raised and fibre particles may go into your eyes. Loose bricks and large pieces of mortar lying between the joists should be removed but leave all the small debris and dust.

Alternatively, if you have decided to use the vermiculite granules, pour it out from the bags and spread it out evenly over the whole floor area until it is just level with the top of the joists. You will have to place a wad of fibreglass or mineral fibre insulation between, and just short of, the ends of each pair of joists under the eaves to prevent the vermiculite flowing under the eaves completely and blocking off the ventilation.

For the second layer, starting farthest from the trap door, lay the 1200mm-wide insulation material across the joists, butting the edges closely together and cutting it so that it fits snugly against the walls and other obstructions, as before. Cover the whole area but, again, stop the insulation where it meets the underside of the rafters, and do not attempt to pack it between the rafters.

There will now be two layers of insulation over the whole ceiling area and if they have been carefully laid there will be few gaps through which heat can rise. A section through the completed insulation is shown in diagram 4.

37

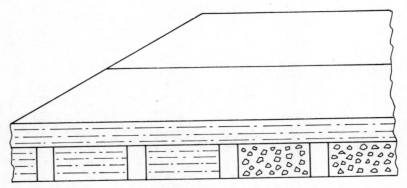

Diag. 4 Section through the completed insulation—new or existing house

TRAP DOOR

For a new house, this should have a simple wooden box surround on its upper side which is filled in with 100mm of insulation. If you can do this on your existing trap door it will make a neat job. Failing this, cut a piece of fibre insulation to fit the top side of the door and secure it in place with wire or string looped round the heads of nails driven into the frame of the door. Clean around the ledge on which the door sits and stick self-adhesive draught stripping around the ledge. In a new house good quality draught stripping should be used.

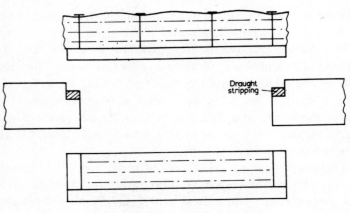

Draught stripping

Diag. 5 Trap door insulation

ALTERNATIVE MATERIALS FOR NEW AND EXISTING HOUSES

Some insulation contractors have schemes whereby mineral fibres or pellets are blown by machine into the loft to the depth required. So far as the insulation is concerned, these are quite satisfactory and your architect will work out the most economical way to insulate your particular loft. However, if this type of insulation is carelessly installed, it will block off the under-eave ventilation and the contractor must not allow this to happen.

7 Insulating an attic or dormer room

Attics are to be found in older houses and the roof construction is, in fact, very similar to newer houses which have had dormer conversions carried out, or indeed a dormer-type bungalow. Attic rooms are notorious for being too cold in the winter and too hot in the summer. As previously explained, the roofing materials are thin and allow heat to escape rapidly in winter. Conversely, in the summer, the tiles or slates become quite hot when the sun is shining on them and heat passes through into the room below, making it uncomfortably hot. Correct insulation will resolve both of these problems.

A cross section through a typical attic or dormer room is shown in diagram 6 and areas which are not floored should be insulated exactly as described previously. In an existing house,

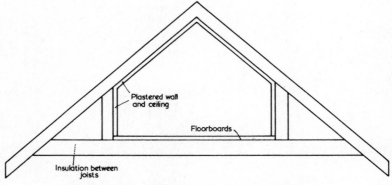

Diag. 6 Section through an attic or dormer room

make sure that all the loft spaces have been dealt with; access to these will be from access doors in the attic or by separate trap doors.

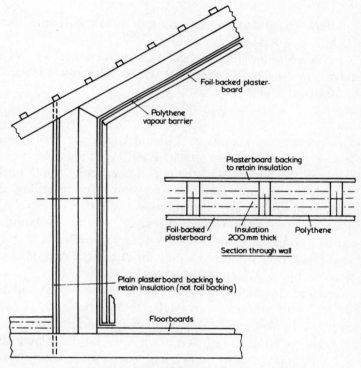

Diag. 7 Detail of insulated ceiling and wall—new house

Dormer room in a new house

The sides and sloping roof are to be insulated by filling in between the timber rafters and the side wall supports with insulation material. The depth of the timberwork will be increased locally to 200mm to give the correct thickness of insulation. The general construction will be seen from diagram 7 and it may well be simpler to use 100 × 50mm rafters, which are then easily increased to a depth of 200mm by nailing to

them lengths of similar timber. The side wall supports could
be constructed similarly to give the required thickness of
insulation.
Suitable materials for use in the construction could be:

Insulation—expanded polystyrene, grade SD with fire-retard-
ing additive (FRA), 50mm thick
Vapour barrier—polythene sheet 125 microns thick
Timber work—should preferably all be impregnated against
rot and insect attack
Panelling—foil-backed plasterboards.

The polystyrene should be fitted between the timberwork
in four layers, each 50mm thick. Each layer must be a snug
fit between the timbers and the end timberwork. Joints in the
layers of insulation should be staggered. By installing the
insulation in layers, any gaps around the edges of one layer
will be covered by a subsequent layer. If the insulation is
installed in one piece 200mm thick it is very difficult to cut it
so that it is a snug fit all round and air currents circulate in the
gaps reducing the effectiveness of the insulation.

Alternative insulation materials are available to the building
trade and your architect may prefer to specify them, in place
of expanded polystyrene. However, this material is about the
most efficient commercial insulation currently available and
it is, in addition, almost completely waterproof. If any water
does come through the roof it will run off the polystyrene,
whereas with some fibre insulation materials it will soak into
the insulation, rendering it virtually useless until the following summer, when it will dry out. There is a slightly increased
fire risk with polystyrene and the addition of the fire-retarding
additive prevents the material from burning of its own
accord. It will only burn so long as flames are playing on it;
as soon as the flame is removed the material extinguishes
itself. One has to be practical about this and realise that if
the fire is of such intensity that it breaks through the plaster-

board to reach the insulation material then it has also reached the timber framework of the roof and the type of insulation used will have little effect on the consequences.

Electric cables which run behind panelled walls really ought to be in a metal conduit, although this is not required by the regulations, and your architect may consider it to be an unnecessary expense.

Attic or dormer room in an existing house

As with a new house the method is to fix wooden battens to the sides and sloping roof structure and to fill in between these with a suitable insulating material. It will finally be panelled over to give a hard surface suitable for decorating. In some cases, it is possible to fix the insulation between the existing timbers and this is also dealt with. The construction is shown in diagrams 8 to 11.

It is not really practicable to fit 150mm of insulation in this case, as the weight and cost of the timber battening becomes excessive. The thickness of the insulation will, therefore, be restricted to 75mm, or possibly 100mm if the insulation is fitted between existing timberwork which is 100mm thick. Naturally this thickness is not up to the general standard we are trying to achieve, but even so the effect of fitting 75mm of insulation right around your attic room will be dramatic. A vapour barrier will be included to prevent condensation inside the panelling.

MATERIALS REQUIRED

Timber battens—rough-sawn 75 × 50mm and preferably pressure-treated against rot and insect attack

Insulation—expanded polystyrene, grade SD with FRA, 38mm thick in sheets 2400 × 1200mm

Vapour barrier—polythene sheeting 125 microns thick and 4 metres wide

Panelling—Supalux, Masterboard, or similar board 9mm thick, sheets 2400 ×1200mm.

The polythene sheeting can be obtained from builders' merchants or some DIY stores and the timber from either a timber merchant or, again, some DIY stores. For the expanded polystyrene and panelling board you will have to go to a board merchant—look up your local Yellow Pages under 'Boards'.

You will have to work out the quantities needed for your own particular attic but as a guide, the materials required to panel an area 2400 × 1200mm would be as follows:

Timber	75 × 50mm	6m
Polystyrene	38mm thick	2 sheets
Polythene	125 micron	1m
Supalux	9mm thick	1 sheet
Nails and screws.		

ALTERNATIVE WAYS OF FIXING THE INSULATION

The insulation can be fixed either between new battens which are fixed over the existing plasterwork or, if there is easy access between the vertical side walls, it can be fitted behind the plasterwork and between the existing upright timber supports. Similarly, if the sloping ceiling is in poor condition it can be pulled down and the insulation fitted between the rafters. This saves on the cost of the new battens but it does mean of course a certain amount of extra mess. You should read carefully through the remainder of this section before deciding which is most suitable for your house.

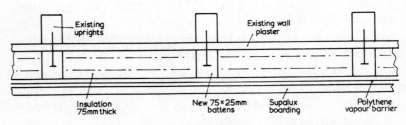

Diag. 8 Section through the insulated side walls—existing house

PANELLING OVER EXISTING PLASTERED WALLS AND CEILING
A section through the completed insulation and panelling is
shown in diagrams 8 and 9.

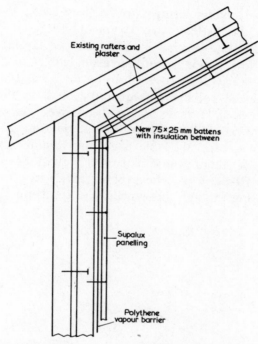

Existing rafters and
plaster

New 75 × 25 mm battens
with insulation between

Supalux
panelling

Polythene
vapour barrier

Diag. 9 Section through the insulated wall and ceiling—existing house

Prick through the wall and ceiling plasterwork to determine
the positions of the existing uprights and rafters; these will
usually be at about 400mm centres. Cut the timber for the
new upright and sloping battens, neatly mitring the joint, as
in diagram 9. Drill screw holes through the battens, about
600mm apart, and screw the battens through the plaster
securely into the timberwork behind, or above, in the case of
the sloping ceiling. You will need fairly long screws, about

100mm long, if you counterbore the holes so that the screw heads are recessed, 150mm long if you do not recess the heads. Nails must *not* be used as the heavy hammer blows needed to drive them in will shake the roof structure and loosen slates or tiles.

Cut the polystyrene, using a fine hacksaw blade, to be a snug fit between the battens; two layers are required. Butt any joints together and overlap joints in the two layers. By using two layers of insulation rather than a single thick one, any gaps in the first layer will be covered up by the second layer.

Open out the polythene sheet and tack it to the battens, trying to cover as large an area with one piece as you can. If a join has to be made, overlap the material and seal the join with polythene tape available from your builders' merchant. An overlap of 50mm should be left all around the edges. This layer of polythene forms the vapour barrier to prevent water vapour passing through the panelling to condense on the cold

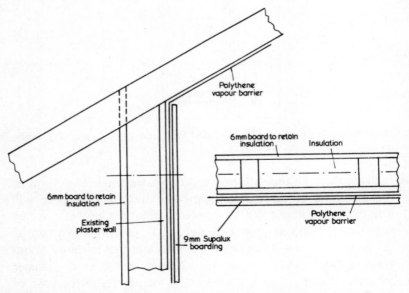

Diag. 10 Insulation fitted between the existing uprights—existing house

original plaster. The sheet need only be loosely tacked as it will be held in place by the panelling, which is fixed next.

Cut the panelling board as necessary and nail it to the new battens using galvanised plasterboard nails. Measure the sheets carefully so as not to waste the material; if the sheets can be fixed in one piece it will reduce the amount of sawing and the number of joints. Any joints must be over the centre of a batten so that the edges of the board can be securely nailed. Use a punch to drive the nail heads flush with the surface of the board.

The excess polythene can now be trimmed off with a sharp knife and all the gaps and joins filled in with a plaster filler. Spot-paint all the nail heads with an oil-based undercoat or primer and seal the board with an oil-based sealer. These boards have a sufficiently smooth surface for most purposes and they can then be decorated as required, but if you want a superior finish then paper the boards with a lining paper.

FITTING INSULATION BEHIND THE SIDE WALLS

If the space behind the vertical side walls is large enough to work in, then there is no need to fix new battens, as the insulation can be installed from behind, between the existing timber supports. The thickness of the insulation must be chosen to suit the depth of these timbers; usually they are 75mm deep, and two layers of insulation should be used. Diagram 10 shows the arrangement.

Again, cut the polystyrene and fit it snugly from behind between the uprights, two layers deep, to bring it flush with the back of the uprights. Older houses will have lath and plaster walls and the back of the laths where the mortar has squeezed out will be very rough. Instead of a first layer of polystyrene, use a layer of glass or mineral fibre insulation material 15 to 25mm thicker than the polystyrene you would have used. This will fill in the rough back and because it will be lightly compressed, it will hold in place when the outer layer of polystyrene is fitted. Panel over the back with 6 or

9mm board to retain the insulation.

A vapour barrier must still be fitted inside the room to prevent condensation within the panelling and polythene sheet must again be tacked over the whole of the wall area inside the room, after you have pricked through the plaster to determine the position of the uprights. Panel over this with 9mm board securely nailed through the plaster into the existing uprights; longer, galvanised, plasterboard nails will be needed for this.

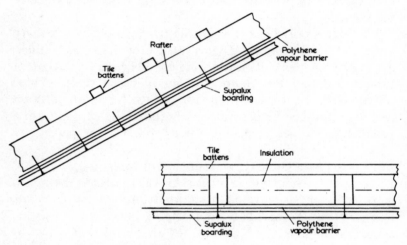

Diag. 11 Insulation fitted between the rafters—existing house

FITTING INSULATION BETWEEN THE RAFTERS

If the sloping ceiling plaster is in poor condition it can be pulled down leaving the rafters exposed. The insulation can now be fitted snugly between the rafters in two layers to bring the insulation level with the underside of the rafters. The total thickness of insulation will be the same as the depth of the rafters. This is shown in diagram 11. Again, tack polythene sheeting over the whole ceiling area to act as a vapour barrier; if one sheet can cover the walls and ceiling in one piece, so much the better, otherwise tape any joins with polythene tape. Panel over with 9mm board, as before.

ALTERNATIVE MATERIALS

The type of boarding suggested provides an adequate finish for most purposes without resorting to plastering. It also protects the timber and insulation from the effects of a small fire. However, if plastering is no problem to you, then plasterboards can be used instead: plasterboard is cheaper but you will have the problem of skimming it over with plaster to obtain a smooth finish. Foil-back boards should be used with the foil side to the wall or roof. Plasterboard used behind the side walls to retain the insulation must *not* be foil-backed as any water vapour inside the panelling must be allowed to escape.

GABLE END WALLS

A lot of attic or dormer rooms will have brick end walls and these may need panelling and insulating as described in chapters 13 & 14. Remember that if you have a semi-detached house and your neighbour is also using his attic or dormer room then there should be no need to insulate the party wall. There will be little heat lost through the wall as the temperature on both sides will be similar—you may even gain some! In a new house the end-walls will be insulated automatically by the cavity wall insulation.

8 Heat losses through windows

There are really only two types of windows that you can have —fixed windows or opening windows. Of the latter there are several varieties; some have side or top hinges, others vertically sliding sashes, horizontally sliding panels, pivoted frames or louvres.

All traditional British windows have one feature in common and that is that they comprise just one sheet of glass. This is effective in keeping out wind and rain but because the glass is so thin it conducts heat rapidly out of the room.

To prevent this loss of heat it is necessary to install additional sheets of glass to trap still spaces of air between them. Air is a good insulator, provided that it is not allowed to circulate, and this is achieved by not making the air space too wide. A comparison can be made by thinking of a river; the water flows faster towards the centre than at the banks because the water at the edges is interfered with by the roughness of and projections from the bank which prevent the water flowing smoothly. Similarly, but on a much smaller scale, the speed of a current of air flowing across glass increases the farther it is away from the glass and at the glass face the air is practically stationary.

Air rises as it is warmed and falls as it is cooled, so that if two sheets of glass are fixed well apart in a window and the inside temperature is higher than that outside, then air currents will circulate inside the window as shown in diagram 12. Heat will be transferred from the inside to the outside via the air.

If the two sheets are brought closer together, then the natural air currents will collide and the circulation is largely prevented (diagram 13). There is then less heat transferred from the inside to the outside. The optimum distance apart for the two panes is 20mm but the space can vary from 10 to 200mm without too much increase in the heat loss. Below 10mm the thickness of the air space is too thin to effectively prevent the flow of heat, while above 200mm the air is beginning to circulate more freely.

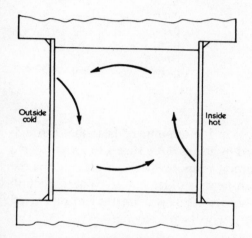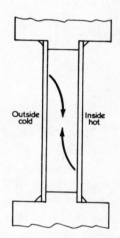

Diag. 12 Air currents circulating inside a window when the panes are well apart

Diag. 13 Air currents colliding when panes are brought closer together

Heat loss figures

The heat loss figure decreases as the number of panes increases, as this increases the total thickness of glass and air (table 11). Comparing these figures with the heat loss figures for walls in chapter 12 you will see that a window having five sheets of glass still loses more heat than an ordinary cavity wall in a newer house. This is simply because the total thickness of glass is still, at most, 5 × 4mm, which is 20mm, compared with the thickness of the brickwork, which is 205mm. The air between the panes is never completely still and so it transmits more heat than does really still air, such as that trapped in insulation materials.

TABLE 11
HEAT LOSS FIGURES THROUGH WINDOWS
HAVING VARIOUS LAYERS OF GLASS

Number of sheets	Number of air spaces	Heat loss figure	Percentage reduction
1	0	0·0057	—
2	1	0·0029	49
3	2	0·002	65
4	3	0·0015	73
5	4	0·0012	79

However, good savings in the amount of heat lost through the windows are obtained by using *three* sheets of glass to give two air spaces. The addition of more sheets of glass raises the cost substantially in both new or older houses. The reduction in heat loss with three sheets of glass compared to one is 65 per cent and this is a very worthwhile saving.

Heat loss calculations
We can work out the amount of heat lost through both single- and triple-glazed windows, the heat loss figures being 0.0057 and 0.002 respectively. The downstairs rooms should be at a higher temperature than the upstairs rooms and an average temperature for the whole of the house of 19°C is used for the calculations, in this case.

By comparing the two figures for your particular size of house you see how much less heat is lost through triple-glazed windows than is lost through ordinary windows. As a point of interest, triple glazing is practically standard in Scandinavian countries.

TABLE 12
AMOUNT OF HEAT LOST THROUGH WINDOWS

	Typical House Size				
	2-bed-roomed terraced	3-bed-roomed semi	4-bed-roomed detached	5-bed-roomed detached	
Window area in square metres	10	15	25	30	
Single glazing Heat loss					
Per hour kW	1	1·6	2·6	3·2	(× 0·0057 × 19)
Per day kW hrs	16	25	41	50	(× 16)
Triple glazing Heat loss					
Per hour kW	0·35	0·6	1·0	1·15	(× 0·002 × 19)
Per day kW hrs	6	10	15	19	(× 16)

Advantages of triple glazing

The benefits of insulated windows are more than just money savings on your fuel bills. One of the greatest advantages is that the cold down-draughts by the window, which can spread across the floor, are eliminated and the area near the window will now be just as warm as the rest of the room.

In living-rooms and bedrooms, condensation on the windows will be almost eliminated and the paintwork will last longer on the windows and the window sills. The tendency for bathroom and kitchen windows to steam up will be reduced, although it cannot be eliminated entirely from these rooms. Good ventilation is the only way to remove an excess of steam.

Triple glazing can also improve your security. A lot of windows can be opened quite easily by the casual intruder and triple glazing combined with good security catches is an excellent deterrent. Because a single sheet of glass is so thin it also allows noise or sound to pass through easily. The addition of two air spaces and two extra sheets of glass will result in a substantial reduction in the amount of noise coming through the windows. This is of particular benefit if you live adjacent to a main road or airport and a section on the best way to reduce noise with triple glazing is included.

9 Triple glazing for a new house

The exact specification for the windows will be worked out by your architect and a summary of the requirements is given here. Usually the two innermost sheets will comprise a sealed unit preferably with the optimum air space of 20mm between the panes. The third sheet will be mounted on the outside and the method of mounting it will depend on the size of the window, as described later in this chapter. Windows are usually broken as a result of balls or stones being thrown or kicked from outside the house and this arrangement puts the cheaper and more easily replaceable single sheet of glass in the line of fire. However, there is no reason why the arrangement should not be reversed and your architect may prefer to fit sealed, double-glazed units in a standard window frame and fit a separate, single-glazed, interior window on the inside, similar to the arrangement shown in diagram 19.

To minimise condensation between the single sheet of glass and the sealed unit a small amount of ventilation is needed between them to the outer atmosphere. This should take the form of small gaps or drilled holes along the bottom edge only of the frame holding the outer sheet or sheets, to connect the space between the single sheet of glass and the sealed unit with the outer atmosphere.

It will occasionally be necessary to clean between the single sheet and the sealed unit and this should be done by cleaning the glass with warm water to which a little detergent has been added. Leather off with clean warm water containing a small amount of disinfectant to prevent mould forming. Let the glass dry before replacing the panes. The glass will stay reasonably clean for two or three years where the sealed unit is on the inside.

Alternatively, the windows can be fitted with triple-glazed, sealed units. The window frames must have sufficiently deep rebates to take these and the final choice will be decided between you and your architect on cost and design.

Fixed windows

For the lowest cost, small windows will have a sealed inner unit and a detachable outer pane. On medium-sized windows the outer pane should be fixed with hinges so that it can be opened easily for cleaning. Typical sections through these arrangements are shown in diagrams 14 and 15.

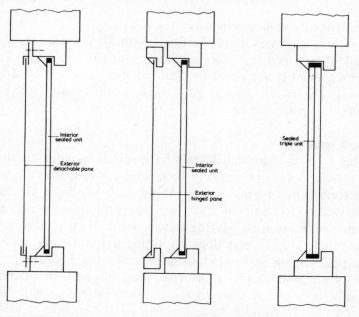

Diag. 14 Triple glazing with detachable exterior pane

Diag. 15 Triple glazing with hinged outer pane

Diag. 16 Triple glazing with a sealed triple unit

55

To provide easy access between the two sets of panes in a large window they should be constructed with a fixed outer pane. The sealed unit inner pane should be divided into two or more sections, each mounted in horizontally sliding frames. Good quality frames must be used with effective draught sealing built into them and diagram 17 shows the arrangement.

Hinged and pivot windows
These will generally be fitted with a double-glazed, sealed unit and a detachable or hinged pane on the outside. Triple-glazed, sealed units can, of course, be fitted if the window frame is suitable. A pivoted window must still be capable of being swung right over with the outer pane in place, for normal window cleaning. In both cases the hinges must be sufficiently strong to carry the extra weight.

Horizontally sliding windows
Two sets of sliders should be fitted, with the inner set carrying sealed units and the outer set fitted with single glazing. This arrangement enables you to clean all the glass surfaces from inside the room. Again, good quality units with adequate draught sealing are needed.

Sash windows
This type of window will only be specified by your architect where he considers they will be in keeping with the style and surroundings of your house. Metal sash windows are now available and some of these will carry double glazing, but none will carry triple glazing and so an auxiliary window must be fitted. This is best done by fitting a proprietary interior vertical sliding sash window fitted with single glazing. The units are not usually counterbalanced but can be operated easily because of their lightness. Your architect will have details of the various makes available. If two sets of sashes are being installed side by side in a larger window then a horizontally sliding unit could be used on the inside.

Patio doors

The amount of heat lost through a double-glazed window is nearly three times that lost through a newer, conventional, cavity wall of the same area. Large patio doors are also a big source of heat loss and it is most important that additional glazing is installed here.

Two sets of sliding units should be fitted, with all units sliding, and no fixed units—see diagram 18—so that all the glass surfaces can be cleaned. The inner set is to be fitted with double-glazed, sealed units and this is a standard arrangement.

How the outer set is glazed will depend largely on whether or not there are children in the house. To give the greatest amount of insulation the outer set will be an exact duplicate of the inner set and will also be fitted with double-glazed,

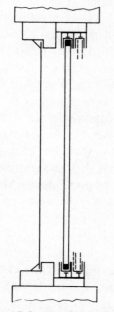

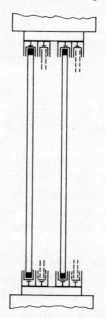

Diag. 17 Large, fixed picture window with horizontally sliding double-glazed inner unit

Diag. 18 Patio doors with two sets of double-glazed sliding doors sealed units. This gives quadruple glazing to these large windows. If, however, you have young children, then it may well be better to glaze the outer set with single, heavy, plate, toughened, or laminated, safety glass which is then more able to stand up to the occasional blow from a tennis ball or football. Your architect will advise you on the most suitable glass to use for your particular door.

General points
Proprietary window frames are available either with an outer, hinged, auxiliary frame to carry the third sheet of glass, the inner rebate being suitable for a double-glazed, sealed unit, or with the rebates suitable for triple-glazed sealed units. Your architect will have details of these. From the insulation point of view the preferred material for the frames is wood because it does not conduct heat so quickly as does aluminium or steel. Clearly there are some types of windows which are just not made with wood frames—for example, patio doors —and in this case, to reduce the heat loss through the metal frames the inner and outer frames should be separated by a small space and must not be in direct contact.

The window frames should incorporate night ventilators. These are small vents in the frame which can be opened or closed as required to allow a small amount of air change in the room without opening the window and they are completely safe to leave open even if you are out of the house. All the opening windows should have good quality, draught-proofing seals built into them.

Sound insulation
The triple-glazing methods described previously in this chapter will considerably reduce the amount of noise coming into your house through the windows. However, if you are particularly troubled by noise, then further reductions in the amount of noise coming through the windows can be made.

All that has to be done is to increase the gap between the

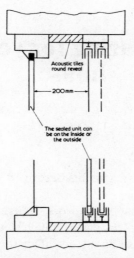

Acoustic tiles
round reveal

200mm

The sealed unit can
be on the inside or
the outside

Diag. 19 Arrangement for maximum sound insulation with a space of
200mm between the two sets of panes

sealed unit and the single sheet of glass, from that recommended, to 200mm. This is the optimum figure for sound reduction. Naturally, two sets of window frames will be required and the wall between the frames should be lined with acoustic tiles. Your architect will have all the technical details required to hand. A section through the window is shown in diagram 19.

10 Triple glazing for your present house

The methods of triple glazing described in this chapter give the desired results for a reasonable cost and they assume that you will be doing most of the work yourself. The existing style of window is retained or, if it is covered with another sheet of glass, is still visible, so as not to affect the outside appearance of your house.

The basis of insulating windows is to trap two air spaces between three sheets of glass and as all windows have one sheet in already, this is retained. Two extra sheets will be added—one inside and one outside the existing glass. There is little point in replacing existing glass with sealed units as you are only having to pay for two sheets of glass when you have one already.

If, however, the existing glass is cracked or badly distorted and needs replacing, then it would be sensible to replace it with a sealed unit. The gap in the unit should be 20mm if possible and you may have to have what is known as a stepped unit to fit into your existing frames. The cost will be greater than for two separate sheets but it will result in a neater installation. The third pane must still be fixed outside the sealed unit to give triple glazing.

Preparation, precautions and checks
Start by completing one small window first and make sure all your measurements are working out correctly before ordering the bulk of the glass. Most glass merchants will cut reason-

able quantities of glass while you wait. Small sheets can be placed on a flat board on the back seat of a car. Place paper between each sheet to prevent them sticking together, protect the seats well and use plenty of packing to stop the sheets sliding about. Larger sheets can be carried flat on a board in the back of an estate car; these floors are rarely flat and the glass may crack if placed directly on the floor. Again, pack the glass well to prevent it sliding about and do not pile the sheets more than four or five layers deep.

Always wear gardening gloves with flexible leather palms when handling glass as bare hands cut all too easily; it is better if the gloves cover your wrists as well. Lift the glass carefully, trying not to slide one sheet on top of another or the sharp cut edge may well score the glass below.

Stack the glass well out of harm's way by leaning it against a wall, with paper between each sheet, and the lower edge resting on two pieces of wood to prevent damage to the edge of the glass and also to make it easier to pick up later. If the glass is outside, ensure that it cannot blow over. Before moving the glass, clear away any bricks or roller skates which might be in your path, pick it up with both gloved hands where possible and move carefully, avoiding obstacles.

Once you have practised on the smallest window and are sure of the sizes, it is preferable to have the rest of the glass delivered by the merchant or DIY store. Transport is then at their risk and not yours and it may well be worth any small extra charge.

TABLE 13
GLASS INSTALLATION

Sheet area in square metres	Up to ½	Up to 1	Up to 1½	Up to 2
Glass thickness	3mm	3mm	4mm	4mm
Working from the floor or ground	1 man	1 man	2 men	2 men
Working from a ladder	1 man	2 men	2 men	

Table 13 shows which thickness of glass to use and when extra help will be needed. If you are in any doubt as to the correct thickness to use, your glass merchant will help you. A sheet of glass any larger than 2 square metres will have to be installed professionally.

Make sure that all the frame timber is sound by trying to push a medium screwdriver into the wood; check the bottom sills especially carefully. If it only makes a small dent, then the timber is probably sound, but if it goes in for any way the timber is rotting and should be replaced. Scrape down and repaint the woodwork—inside and out—replacing any missing mastic or putty as the extra glazing will be on for a few years before it needs removing to clean the glass faces again.

Fixed-pane timber-frame windows
These are the easiest to do first. The inside pane will have a simple flexible plastic channel fitted around the edges, both to protect the sharp edges of the glass and to form a seal between the glass and the window frame. The panel is then held to the frame with small plastic clips screwed into the timber.

It is cheapest to fix the outside sheet in the same way, but it is better mounted in a rigid plastic or aluminium frame which is then screwed to the window frame. This is slightly more expensive but gives a much neater appearance.

MATERIALS REQUIRED

Firstly and obviously the glass: this should be float glass for clarity and a guide to the thickness needed has been given in table 13. It can be purchased from glass merchants or some DIY stores. There are many makes of the flexible plastic channel for interior use and you should purchase it complete with the clips and screws. It can be purchased from virtually any glass merchant, DIY store and some multiple stores. Exterior rigid channel for 3mm glass and aluminium channel complete with a plastic liner for 4mm glass, can be obtained by mail order from the firms listed in the Appendix. You may find alternative exterior glazing channel in your local DIY store or glass merchants and you should check the price and quality before deciding which to use.

A section through a typical window is shown in diagram 20 and it shows the measurements to make for the inside pane of glass; add 15mm to the dimension indicated. Use a steel tape which is in good condition and try to use the metric markings, as glass merchants now work in metric dimensions. With a little practice you will soon become used to the system. For the exterior panes, follow the instructions included with the channel, as the size of the glass required varies slightly from one manufacturer's system to another. Measure the sizes both vertically and horizontally, write them down and *recheck* the measurements. The length of channelling needed should be rounded up to allow for cutting.

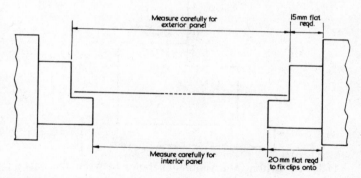

Diag. 20 Measurements to make for the interior and exterior glazing

INSIDE PANEL

To fit the clips and plastic channel, a flat face 20mm wide is needed around the window frame as shown in diagram 20. If necessary, screw suitable lengths of planed wood to the inside of the frame to give a flat face of this width.

If any moist air from the room seeps into the space between

the panes, condensation may take place on the existing pane and, to reduce the likelihood of it happening, we must provide a means of escape for the water vapour before it can condense. In this respect we are lucky, as water vapour will transfer itself from air having a higher moisture content to air having a lower content, so all that is needed is a small amount of ventilation from between the panes to the outer atmosphere.

In the lower window frame only, drill 6mm holes at 300mm intervals to connect the two air spaces, as in diagram 21. Blow through the holes with a car or bicycle pump to make sure they are clear. These are quite sufficient to ventilate the inner-most air space so long as the inner panel is reasonably sealed.

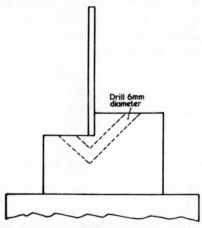

Diag. 21 Drilling the ventilation holes between the panes

Press the flexible channel around the perimeter of the glass and neatly mitre the corners by cutting through the sides only of the channel with wire cutters or strong toe-nail scissors. Finally butt the two ends neatly together. Next, clean both the existing window and the new panel with warm water to which a little detergent has been added. Leather off with clean warm water containing a little disinfectant and let the glass dry.

The inner panel can now be secured to the window frame

with the plastic clips at 200mm intervals all around its edge. Do not screw the clips too tightly—just sufficient to nip the plastic channel against the frame. If they are too tight you will risk breaking the glass—and with glass the price that it is, that is something to be avoided at all costs.

EXTERIOR PANEL

The surround to be used with 3mm glass is a rigid plastic channel into which the glass fits. The built-up panel is screwed to the window frame through the channel as in diagram 22. Cut the channel using a Junior hacksaw, which has fine teeth, and mitre the corners carefully to give a neat finish. It is better to cut the channel for each side a little too long to start off with and it can then be trimmed off as you work your way round the glass. It is difficult to add on to a piece which has been cut too short!

Follow the instructions supplied with the channel and drill small ventilation holes through the bottom channel only to ventilate the space between the outer panel and the existing glass. Clean the glass faces as previously described and secure the panel to the frame with either the aluminium screws provided or small, round-head, brass screws.

The surround for 4mm glass is made from aluminium channel and it has a U-shaped flexible plastic liner to grip the glass. Again, cut the channel and mitre the corners neatly, fitting the plastic liner all round the glass before pressing on the aluminium surround. Clean the glass and again secure the panel with brass screws.

Both these channels only need a flat face 15mm wide to attach them on. If you are unlucky enough not to have such a facing around the window frame then again secure lengths of wood round the window frame to give a suitable facing. Make sure the wood is well painted before attaching the exterior panel. A section through the completed window with both interior and exterior panels attached is shown in diagram 22.

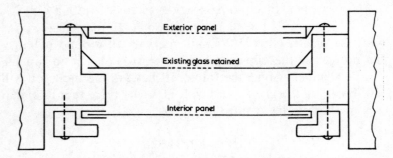

Diag. 22 Section through the finished triple-glazed window

Georgian windows with small panes

It is better to glaze over these with one or more large sheets of glass to suit the frame. This will make subsequent routine window cleaning much easier. Do not attempt to do each small pane separately as there will be far too much work involved and the end result would look somewhat messy. A section through the finished window would be similar to that shown in diagram 23.

Wooden side or top-hung opening windows

Carefully check the timber work as before, paying particular attention to the hinges. Make sure all the screws are in and are tight. If necessary, cut a fresh rebate for the hinges and move them to get the screws gripping properly. The hinges should be replaced if they are old and slack.

To avoid straining the hinges, *only one* additional panel should be fixed to opening windows; two extra sheets of glass would make the window too heavy. If the inside of the window has the necessary clearance when shut and the handle or catch can be operated without fouling the extra panel, then fit an internal panel as described previously. If the internal panel is going to prove difficult to fit, then fix an exterior panel on the outside instead. Make sure that the exterior panel does not catch on the brickwork when the window is opened; you may have to fix a stop to prevent the window from being opened too far.

66

Wooden pivot windows

Again, on existing windows it is better to attach only one additional panel of glass to avoid straining the hinges. In most cases the panel will have to be fitted to the outside because the handles and catches on the inside are usually in the way. Most pivoted windows can be swung right over which allows you to work easily from inside the room when fitting the exterior panel.

Metal-framed windows

Older types of metal windows have several small panes in the frame, rather like Georgian windows, and the additional panels should again be made in one piece where possible to cover the whole of the fixed window. Some metal windows are fixed into a wooden surround and the interior and exterior panels can be easily secured to the surround as described for a fixed wooden window. This is shown in diagram 23.

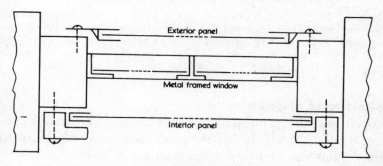

Diag. 23 Metal-framed window fitted with triple glazing

If, however, the metal frame is fixed directly to the brickwork then a wood surround must be made and fixed around the frame —screwed into the wall and not the metal. Do this both inside and outside to allow the extra panels to be attached. The glass in a metal window goes right into the corner of the frame and there is no room for securing screws. Choose the timber size to give the correct air space; fill all the gaps with putty

and paint the new timber well before attaching the new panels.
Fit the wood surrounds slightly clear of the metal frame so
as not to block the ventilation holes, as shown in diagram 25.
Again, ventilation holes must be drilled under the lower edge
of the metal frame to connect the two air spaces. If the metal
frame is set into the brickwork you will need to use a masonry
drill to do this.

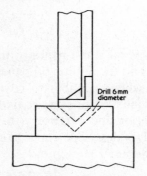

Diag. 24 Metal-framed
window with wood surround
showing the drilling for
ventilation under the metal
frame

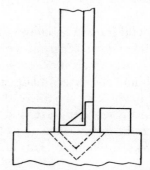

Diag. 25 Metal-framed
window with new wood sur-
rounds fitted to carry the
exterior and interior panels

Steel-framed hinged windows

Again, only one extra panel will be fitted and the metal frames
are usually shaped so that the extra panel can only be fitted
on the outside. Make up an exterior panel to suit the size of
the window, noting that it may have to be made slightly
shorter than the height of the window to clear some types of
protruding hinges. Alternatively, it may be possible just to cut
a piece out of the plastic or aluminium channel surround to
clear the hinge.

If the panel were mounted straight on to the metal window,
the air space would be too narrow, and to correct this a bead
of planed wood, about 15 × 19mm should be fitted between
the panel surround and the metal frame. Bolt the wood to the

68

frame and paint it well. Clean the glass and secure the exterior panel with brass or rust-proofed bolts, passing through the surround, wood bead and metal frame. The bolts must not be too long or they will prevent the window from closing properly. The arrangement is shown in diagram 26.

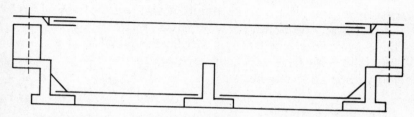

Diag. 26 Hinged metal window fitted with an exterior panel

Horizontally sliding metal windows

These are found in newer houses and there is no cheap solution to insulating these windows. A further complete sliding frame carrying two double-glazed, sealed units will have to be mounted on the inside of the existing window.

If you can do this yourself then suitable frames can be bought from glass merchants and some DIY stores. Choose the best you can afford making sure that it has built-in draught seals. Square the frame up carefully when fitting it to ensure that the panels slide easily. Again it is well worth looking through the *Which?* reports on these systems. It could even be worth your while having it professionally fitted by a reputable company specialising in this type of product.

Louvre windows

Existing louvre windows will not carry double glazing and if there are just one or two louvres at the top of a fixed window, and you still wish to use them, fit an interior sliding unit over them. You could make this up yourself with parts from the DIY shop and it need only carry single glazing but you may have to build it out on a small subframe so that it clears the louvres and the operating handle. The remainder of the fixed window should of course be insulated by fitting both an exterior and an interior panel.

A full set of louvres, running the full depth of the window, will have to have an auxiliary window fitted on the inside. This could be a proprietary wooden-hinged window, opening inwards of course, complete with its frame, if you can find one of the correct size. Alternatively you could make up your own hinged or slider unit with parts from the DIY store; a wood surround may have to be made to carry these clear of the louvre operating handles.

Large picture windows

The size of the glass sheets needed to triple glaze these will be far too large for handling on a do-it-yourself basis. The windows will need to be fitted with interior sliding windows and this should be done by a reputable company. Two, three or even four sliding units may be needed, depending on the size of the window and they should use float glass for clarity. The arrangement will be very similar to that shown in diagram 17. There are several nationally advertised companies making these and you should obtain competitive quotations to see which is best for your particular house.

Vertically sliding, wooden, sash windows

These sliding windows, suspended and balanced by ropes and weights, are common in older houses and a frequent complaint is that they rattle in high winds and are draughty. Some windows were initially fitted with metal chains or braid and these can have a very long life but the rope cords which are usually fitted are a constant source of trouble. So long as the window timberwork is generally sound most of these defects can be eliminated as a result of the insulation.

The top of a sash window is close to the ceiling and when both sashes are opened to overlap each other the hot air at the top of the room spills out rapidly to be replaced by cold air coming in through the lower sash. With only the lower sash open there is still a gap between the two sashes which allows the air near the ceiling to escape, but it does so more slowly. A lot of heat is lost if the top sash is opened and, for all practical purposes, adequate ventilation can be obtained by using the lower sash only. This allows the upper sash to be permanently fixed in position, which a lot are in any case, and it is a good idea to fit a door security chain to the lower sash which allows it to be left slightly open with the knowledge that this will deter the casual intruder. Again, if the room has two sets of sashes side by side it is usually sufficient if only one lower sash is kept in full working order. This allows the remaining sashes to be completely sealed.

Fitting the extra panels with both sashes fixed

Close both sashes fully and screw three lengths of 38 × 19mm planed wood into the inside tracks above the lower sash as shown in diagram 27. The two vertical pieces may need shaping at their lower end to blend in with the sides of the lower sash if these are not square. Along with the lower sash

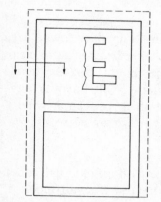

Diag. 27 38 × 19mm wood screwed into the inside track (looking from inside the room) — vertically sliding, wooden, sash window

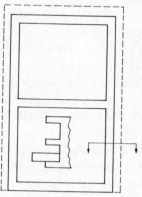

Diag. 28 38 × 19mm wood screwed into the outside track (looking from outside)—vertically sliding, wooden, sash window

these will now form a continuous rebate round the window to which the internal panel or panels can be fixed; they also secure the window.

Similarly, screw two pieces of wood into the sides of the outer tracks, level with the outside of the upper sash, and a piece of wood 32mm square screwed along the sill in line with these will complete a rebate on the outside to which the external panel or panels can be attached. Now remove the lifting handles from inside, putty up any gaps and paint all the woodwork well, working paint into the sash sides to seal them in the tracks.

Make up an internal panel and clip it to the inside rebate, just formed. There is no need to drill any ventilation holes because the gap between the two sashes where they meet serves this purpose. If the panel is going to be more than 150cm high, make it in two pieces using the upper horizontal frame of the lower sash on which to make the joint between the two panels. The lower panel will now need to be ventilated by drilling 6mm holes through the bottom of the lower sash, but the upper panel will be ventilated again by the existing gap between the two sashes.

Similarly, make up one or two exterior panels; these will need ventilation holes drilled in them along their lower edge only. Secure them to the outside rebate, remembering to clean all the surfaces of the glass first. A panel covering the whole window, and which has a plastic edging, will have to have strips of wood packing—about 3mm thick by 15mm wide— placed under its outer edges so that it sits down correctly.

Sliding sashes

Fix the upper sash in place and glaze with an external panel. Ensure that the sash is right up and screw a small block of wood under each side in the track to secure it in place. Paint the sash, again working paint into the sash sides to seal it in the track, and attach an exterior panel to it, exactly as for a fixed window.

The lower sash is retained and, to take the weight of the internal panel which will be fitted to it, the ropes and counterbalance weights must be altered. A metal chain will be used and once fitted, it should have a life of twenty years or more. Extra materials required are three $\frac{1}{2}$in pitch $\times$ $\frac{3}{32}$in bicycle chains, a length of scrap lead waste pipe and a tin of Filtrate 'Linklyfe' chain grease. The cycle chain can be bought from a cycle shop and the chain grease from a motor cycle dealer; your local plumber will probably give you a short length of old lead pipe.

A short length of chain will be permanently screwed to the sash on each side, and will be connected by a spring link to a longer length running over the pulley to the counterbalance weight. This arrangement makes the chain easy to disconnect if the sash has to be removed later. A standard cycle chain has 116 links and is 58in long. It will form the longer chain. The shorter chains are obtained by cutting the third chain to give the two lengths required and the cycle shop will do this for you with their chain cutter. For sashes up to 900mm high the two shorter chains should have 17 links, and for sashes over 900mm high they should have 27 links. Heat the chain lubricant and immerse the chains, as instructed on the tin. This will lubricate the chains for life; the grease on the outside of the chains should be left on to prevent them from rusting.

Once you have worked on a sash window you will realise how simply they are constructed and a brief description of how to remove and refit the sash is included here; if you have any doubts, have a word with a friend who has replaced his own rope sash cords.

Close the lower sash and cut the two supporting ropes, letting the weights drop into the boxes on either side of the window. Carefully prise out the two side beads, starting from the centre—the ends are mitred in—and the sash will now be free and can be lifted out.

Remove the screw or screws in the track, securing the cover

on the weight boxes; some are dovetailed in at the upper end and should be prised out with a down-and-away movement. At the same time, the centre parting, or guide strip, will probably come away at the bottom.

Remove the iron weight from inside the box and check that you have got the correct weight. If the upper sash cord is broken there will be two weights in the box, the larger of which is usually for the lower sash. If there are two sets of sashes side by side, the centre box could contain up to four weights. Remove the rope from the sash sides and the weights.

Remove the two pulley wheels from the tracks. These are in a metal housing held in by two screws which usually have been painted over. Prise out the metal housing and check the axle for wear. If it is worn and slops around, allowing the pulley to bind against the sides of the casing, buy two new pulleys from a good plumbers' merchant or ironmongers. Check that the cycle chain will pass over the pulley freely; if necessary, file the casing with a small square file to clear the chain.

Remove the paint build-up from the sides of the sash with a rasp and the paint in the tracks with a scraper. Sandpaper the sash, track and inner edge of the bottom of the upper sash which is now exposed. Paint the sash and frame with undercoat but do not paint the sides of the sash which are already unpainted.

Screw the shorter lengths of chain into the grooves on either side of the sash with 1in countersunk screws, 6 each side for smaller sashes and 8 each side for larger sashes. The top of the chain should be 100mm below the top of the sash side and it may be necessary to lightly chisel out the groove to ensure that the chain is flush with the sash edge. A sketch of the arrangement is shown in diagram 29.

To work out the extra weight to be added to *each* counter-weight, multiply the area of the new internal pane by 4, for 3mm glass, or by 5 for 4mm glass. For example, if the new pane is 800mm by 900mm, which is 0.72sq m, and 3mm glass

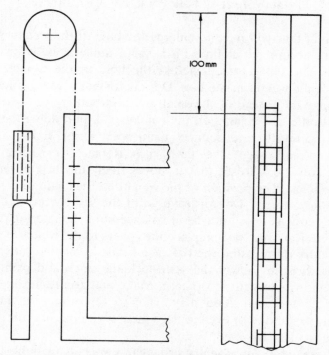

Diag. 29 Arrangement of the counterbalance weight and chain for the lower sash—vertically sliding, wooden, sash window

is to be used, then the weight to be added to *each* side is 0.72 × 4, which is 2.9kg. If the sash lifts correctly then this will be about right. If, however, it is heavy to lift and tends to fall back, then add another ½kg to each side. Conversely, if the sash tends to go up on its own accord, reduce this amount by ½kg per side. Now cut two pieces of lead pipe to the correct weight.

Pass the long chain over one of the pulleys and relocate the pulley in the track. Feed the chain over the pulley into the box and stop the chain from falling right in by putting a nail through the last link and against the pulley. Refix the pulley, securing the screws. Pull the lower end of the chain out of the weight box and thread the lead weight up it. Pass the chain through the hole in the cast iron weight until the top of the weight is up to the top of the weight box opening. You

75

may find that you have to enlarge the hole in the weight with a drill. Secure the chain with a small nut and bolt through one of the links, to jam across the hole in the weight and replace the weight in the box. Do this for both sides. A spring link may do instead of the small nut and bolt.

With the sash resting on the window sill, pull down the top chains carefully and connect them with spring links to the chains already secured to the sashes. Replace the sash in the track and check to see that it moves freely up the full length of the track. The position of the weight on the chain may need slight adjustment. Do not leave go of the sash because it will rise rapidly as it will not be in balance until the interior panel is fitted. Raise the sash, replace the covers to the weight boxes, the centre parting and the two side beads by bending them to spring them in. Lower the sash, lock the catch and paint the sash and frame with a top coat. Make and secure an internal panel to the sash. You may have to reposition the lifting handle or handles or even replace them with ones of a different shape to clear the inner pane.

Alternatively, proprietary sash chains are still available from some builders' merchants or ironmongers and these have rust-proofed links with brass rivets. Two grades are available, one for sashes up to 30kg and one for sashes up to 50kg, total weight. If you can find a local stockist you may prefer to use them in place of the cycle chain. Good quality pulleys with nylon or brass wheels are also made and if it is necessary to replace the old pulleys it is well worth using these. A manufacturer of these items is listed on p 144.

Sound insulation
The triple and double-glazing methods described, coupled with the draught proofing described in the next section will con-siderably reduce the amount of outside noise coming in to your house through the windows. If, however, you live on a par-ticularly noisy road or near to an airport then it is worth altering the arrangement slightly.

All that is needed is an increased air gap between the panes. The maximum reduction in noise is obtained with a 200mm air gap but most window sills are not so wide and a gap of 75 or 100mm is usually sufficient and is easier to install. The arrangement is shown in diagram 30, and you will see that the exterior glazing is done exactly as previously described. The inner panels are mounted on a separate wood frame screwed into the window reveal to give the required air space between the

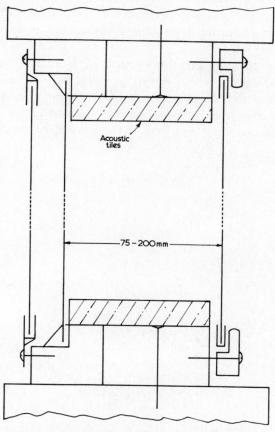

Diag. 30 Arrangement of window panes to give maximum sound reduction

existing pane of glass and the inner pane. If there is an opening window you will have to purchase a hinged or sliding window kit for the inner panels so that you can reach the existing window.

Stick acoustic tiles, which can be obtained from better DIY stores or glass merchants, all around the reveal between the existing glass and the new interior glass. Sound waves spread out as they come through the glass and bounce off the hard wood reveals. Acoustic tiles are made from relatively soft fibrous material and they partly absorb and break up the sound waves giving a further reduction in the amount of noise coming through.

You will be quite surprised at how effective this is in cutting down troublesome noise. A further point worth considering is that if you live close to an airport grants are sometimes available to enable you to insulate your windows and it is well worth checking with your local council before you start.

11 Draught proofing and ventilation

Ventilation in a house is essential to supply fresh air for the occupants to breathe, to remove odours and steam, to provide air for the burning of fuels, such as gas, oil or coal and for the removal of flue gases. A happy balance is needed to ensure freshness without undue waste of heat. Open fires, for example, require large quantities of air which all has to be heated as it passes through the house and is then just lost up the chimney.

Draughts

Doors and opening windows all have gaps around them when closed, simply because it is not possible to manufacture them with an absolutely tight fit. As the house becomes older they warp, the hinges sag or the house foundations settle slightly, and larger gaps are the result. Gaps around the doors and windows can be quite easily up to 2mm or more in width. If we measure up the distance around all the external doors and opening windows and work out the total area of all these gaps the surprising results seen in table 14 emerge.

In addition each chimney flue for an open fire is about the same area as two bricks—again, wide open to the atmosphere. In a warm house air rises up the chimney flues and this draws cold air into the house even if the fires are unlit.

Heat loss

To keep fresh a room which is occupied, the air needs to be changed twice every hour. Unoccupied rooms do not need the air changing and, as an average, it can be taken that all the air in the house needs to be changed once every hour. The volume of the house can be measured and each cubic metre of air requires 0.00036kW to heat it up 1°C. The average temperature in the house is 19°C. The amount of air entering a house which has no draught proofing is easily twice the amount

TABLE 14
GAPS AROUND DOORS AND WINDOWS

	Typical house size			
	2-bed-roomed terraced	*3-bed-roomed semi*	*4-bed-roomed detached*	*5-bed-roomed detached*
Distance around the doors and windows: *cm*	2000	2500	3000	3500
Area with a 2mm gap all around in *sq cm*	400	500	600	700

TABLE 15
COMPARISON OF HEAT LOSSES IN DRAUGHTY AND DRAUGHT-PROOFED HOUSES

	Typical House Size				
	2-bed-roomed terraced	*3-bed-roomed semi*	*4-bed-roomed detached*	*5-bed-roomed detached*	
Volume cu m	150	200	315	450	
Heat losses for a draughty house					
Air change every hour cu m	300	400	630	900	
Heat loss					
Per hour kw	2·0	2·6	4·2	6·0	($\times 0\cdot00036 \times 19$)
Per day kW hrs	32	40	66	96	($\times 16$)
Reduced losses for the draught-proofed house					
Air change every hour cu m	150	200	315	450	
Heat loss					
Per hour kW	1·0	1·3	2·1	3·0	($\times 0\cdot00036 \times 19$)
Per day kW hrs	16	20	33	48	($\times 16$)

required for necessary ventilation and we can work out the comparisons shown in table 15.

By comparing the two tables you will clearly see the cinsiderable saving in heat resulting from the prevention of unnecessary draughts. The figure shown under the loss per hour is in fact the size of electric fire needed just to warm the air which is coming into the house on a cold day. A well draught-proofed house will obviously be more comfortable to sit in than will be a draughty house.

Sealing Materials

All manner of draught-sealing devices are available, some cheap and some expensive. Practically all are effective in that all they have to do is to fill a small gap and be flexible to some extent. For a new house the door and window frames can sometimes be obtained with factory fitted seals and these should be used wherever possible. However, it should be borne in mind that wooden doors and windows require painting every few years and paint can damage and affect the operation of some types of seals. For existing houses, the basis of selection is therefore to use the cheapest product which will do the job effectively, is easily fixed and easily removed for redecoration so that it can be renewed afterwards.

Self-adhesive, smooth-coated, plastic foam strip meets all these requirements. It can be obtained from virtually any DIY store or multiple store. The uncoated variety is cheaper but becomes dirty quickly because it cannot be wiped clean. A seal for the bottoms of the doors will also be needed.

Doors

Only the following doors should be draught proofed: all exterior doors, the cellar head door and the attic staircase door. To allow air to circulate inside the house, internal doors should not be draught proofed.

Clean the rebates in the frame and stick the foam stripping to the top and sides of the frame as shown in diagram 31.

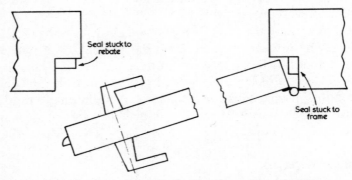

Diag. 31 Draught stripping fitted on a door frame

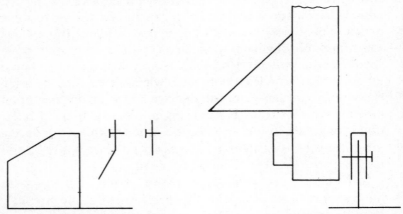

Diag. 32 Draught excluder
for an external door

Diag. 33 Draught excluder
for an internal door

If the door has no threshold, fit a draught excluder of the
type shown in diagram 33 to the bottom of the door. There are
so many of these on the market that it is not possible to
recommend a particular make. A self-lifting type may be
needed if the door opens over a carpet. Exterior doors with the
threshold in good condition should be fitted with a simple
seal fixed along the bottom of the door where it meets the
threshold, as in diagram 32. If the threshold is in poor condition or worn, fit a new threshold and seal. Some modern metal

thresholds already incorporate a built-in seal. The letter box should be lightly sprung or weighted to hold it closed against the wind. If it flaps open replace it with a larger modern one.

If, in a new house, the door frames already incorporate draught stripping, then so much the better. If they do not, then the exterior doors should be fitted with draught stripping, exactly as above. Metal thresholds are generally superior to wooden ones and they should include a built-in seal. Make sure the letter box is a sensible size—larger rather than smaller.

Opening windows

A lot of modern window frames already incorporate seals and this type of window should be specified for a new house, if the cost permits. Sliding window frames should include effective and durable draught stripping. On wood-hinged windows, clean the window frame and stick the seal in place in the rebate against which the window closes. On the hinge side it should be stuck to the frame side, as in diagram 31. Metal windows should have the seal stuck in the rebate all round as in diagram 34.

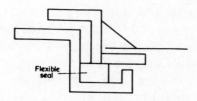

Diag. 34 Draught sealing a metal-hinged window

Pivoted windows should, again, have the seal stuck to the rebate. If the windows are fairly new they may already have seals fitted. On sash windows, only the remaining, opening, lower sash needs any attention. A strip of flexible plastic seal should be tacked along the top of the sash to seal the gap between the sashes when they are closed, as in diagram 35.

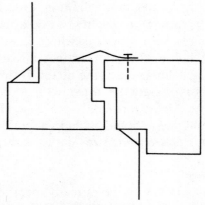

Diag. 35 Draught proofing the gap between sash windows

Stick a length of self-adhesive foam strip to the underside of the sash but leave the sides alone to provide a small amount of ventilation.

Floors

The floors of a new house will have been effectively draught proofed as they are laid by the insulation techniques described in chapter 16. Concrete or solid floors, of course, need no attention as regards draughts. The timber-suspended floors of an older house sag slightly and move with age to create gaps between the floor and the skirting-boards and between the individual planks. Once the floor has been insulated as described in chapter 17, any such gaps will have been covered.

As a stopgap, quarter-round beading should be tacked to the floor all around the periphery of the room hard up against the skirting-boards. It must be nailed to the floor only, not the skirting-board, to maintain the seal as the floor rises and falls as you walk over it. This is shown in diagram 36.

Chimneys and grates—new houses

To prevent solid-fuel fires drawing an excessive amount of air from the room it is imperative that grates draw their air

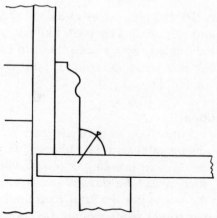

Diag. 36 Quarter-round beading nailed to the floor all around the room

supply from beneath the floor as in the 'Baxi' type of grate. There are several similar designs available. The chimney should be fitted with a draught restrictor which can be closed when the grate is not in use. Failing this, a fitted fire-screen should be made, as described for existing houses below, to prevent wind blowing down or up the chimney.

Solid fuel fires should preferably be the totally enclosed high-efficiency type, as they are more economical than an open fire. If a gas fire is fitted as a permanent installation, the chimney pot should be an air-restricting type with a cover to prevent rain coming into the chimney.

Chimneys and grates—existing houses
Close any fire-place opening not in use with a piece of 19mm block or chipboard. Cut the board 3mm smaller all round than the opening in the grate and tack draught stripping all around the edges. Drill six 15mm holes through the board to provide a small amount of ventilation and press the board into the opening. Decorate it to suit the room. Naturally a board must *not* be fitted if there is a gas fire fitted to the grate. If a chimney is going to be left permanently unused, or

indeed if a gas fire is going to be fitted, it should be swept clear of old soot deposits. The old chimney pot should be replaced by a ventilated cap to keep the rain out and reduce the amount of air passing up the chimney; this is a job for your local plumber or builder.

A word of caution

Gas and solid fuel fires need to draw air from the room so that the fumes pass away safely up the chimney. If you make too good a job of the draught proofing in rooms which have these fires in them, then it can be dangerous. Normally they will draw air from the house but the draught caused can be irritating especially if it comes from around the door into the room and across the floor. The solution is to omit some draught stripping from a window near the fire or, in a ground-floor room, to provide a draught for the fire which is not a nuisance, to drill holes through the floorboards, one either side of the grate.

Gas and solid fuel boilers in the house must be ventilated properly and there will be a grating which allows the correct amount of air to reach the boiler. This must not be blocked in any way.

12 Heat losses through external walls

Most British house external walls are built in two layers: an outer leaf of stone, brick, concrete or artificial stone, and an inner leaf of brick, breeze or concrete blocks. The blocks in recently built houses are aerated to provide a measure of insulation to comply with the 1976 Building Regulations.

Prior to the turn of the century, the two leaves were built close together with whole stones joining them at intervals to make a solid wall. This method of construction is strong but if there are any defects in the wall, then rain can soak through and damage the internal plasterwork and decorations.

Since the early 1900s, the two leaves have been separated by a space of about 50mm and metal connecting links give rigidity to the wall. Any water penetrating the outer leaf now runs down inside the wall to soak away at foundation level. The inner wall and plasterwork remain quite dry. Sometimes this cavity is ventilated by building air bricks into the wall.

Older walls were built using mortar, and the vertical joints in the stone or brickwork are sometimes found to be loose or crumbling and driving rain can penetrate between the joints.

Stone, concrete and brick all conduct heat relatively easily and neither cavity nor solid walls on their own prevent the flow of heat from inside to the outer atmosphere. The walls usually present the largest surface area through which heat is lost.

To rectify this, it is necessary to increase the thickness of the external walls by using insulating materials which contain a large amount of trapped air. In new houses the insulation will be incorporated within the wall cavity and the materials used are either water-repellent or water-resistant so that any rain driving through the outer leaf does not pass through the insulation. In existing houses a layer of insulation will be fixed to the inside of the external walls and panelled over to provide a hard durable surface. In addition, if you have cavity walls then insulation will be installed inside these as well.

Heat loss figures

The heat loss figure for a typical older cavity wall with a brick inner leaf is about 0.0016. For a newer house it is 0.001 because aerated concrete blocks, sometimes called insulation blocks, will have been used for the inner leaf instead. An old house with solid walls will have a higher figure—about 0.002. The effect of adding insulation to these walls is to decrease the heat loss figure, as the thickness of insulation increases. The values achieved for various thicknesses of insulation are shown below.

Initially different wall constructions have a wide variation in the rate at which they lose heat; a solid wall loses heat twice as quickly as one built with insulation blocks on the inside.

TABLE 16
HEAT LOSS FIGURES FOR EXTERNAL WALLS
WITH DIFFERENT THICKNESSES OF INSULATION

Insulation thickness mm	Solid wall	Cavity wall—inner leaf brick	Reduction in loss %	Cavity wall— inner leaf insulation blocks	Reduction in loss %
Nil	0·002	0·0016	—	0·001	—
25	0·0009	0·0008	50	0·0006	40
50	0·00055	0·0005	68	0·00045	55
75	0·0004	0·0004	75	0·0004	60
100	0·0003	0·0003	81	0·0003	70
150	—	—	—	0·0002	80

TABLE 17
AMOUNT OF HEAT LOST THROUGH THE EXTERNAL WALLS

	Typical House Size				
	2-bed-roomed terraced	3-bed-roomed semi	4-bed-roomed detached	5-bed-roomed detached	
Total wall area sq m	50	75	135	160	
Older cavity walled house—inner leaf brick					
Heat loss					
Per hour kW	1·5	2·1	4·0	4·8	(× 0·0016 × 19)
Per day kW hrs	24	35	62	77	(× 16)
Newer cavity walled house—inner leaf aerated concrete blocks					
Heat loss					
Per hour kW	0·9	1·4	2·5	3·0	(× 0·001 × 19)
Per day kW hrs	14	22	40	48	(× 16)
Existing house with 100 mm of insulation					
Heat loss					
Per hour kW	0·3	0·45	0·8	1·0	(× 0·0003 × 19)
Per day kW hrs	5	7	13	16	(× 16)
New house with 150 mm of insulation					
Heat loss					
Per hour kW	0·2	0·3	0·5	0·6	(× 0·0002 × 19)
Per day kW hrs	3	5	8	10	(× 16)

However, as the insulation is added, it becomes so effective that the construction of the wall makes no difference to the heat loss figure.

For a new house a very high standard of insulation should be aimed at and the thickness of the wall insulation will be 150mm. For an existing house, 100mm of insulation will be aimed for. It is unreasonable to go for a higher figure simply because the size of the battening needed for the panelling would become excessive. Even so, a reduction in the rate at which heat is lost through the external walls of over 80 per cent is obtained.

Heat loss calculations

We now have all the information to hand to enable us to work out how much heat is lost through the external walls, and the calculations are very similar to those contained in the previous sections. Because the walls surround both the upstairs and downstairs rooms, an average temperature difference of 19°C is used.

Figures from table 17 have been used in the summary in chapter 3 to work out the total amount of heat lost from the house.

By looking down this table for your particular size of house you will see the extent to which the heat lost through the walls decreases as the thickness of the insulation, and its consequent effectiveness increase.

13 Insulating the external walls of a new house

The exterior cavity walls will be insulated by incorporating an insulating material into the cavity. The walls will otherwise be built using standard materials but the width of the cavity is to be increased as specified below to give the necessary thickness of insulation.

A section through the completed wall is shown in diagram 37. The outer leaf can be any standard material to suit the style and appearance of the house. The wall cavity is to be 150mm wide. Wall ties 255mm long will be needed to tie the two leaves together and care must be taken to ensure that mortar does not fall into the cavity as the building work progresses. The inner leaf should be aerated concrete blocks (insulation blocks) having a minimum thickness of 100mm. Ventilation bricks for ventilating timber-suspended ground floors should be sleeved during construction to prevent them from becoming blocked by the insulating material.

Insulating the walls
There are two ways in which this can be done: either slabs of insulating material can be placed between the two leaves as they are being built or the insulation can be injected into the cavity after the walls have been completed.

The slabs of insulation are made from mineral or glass fibre

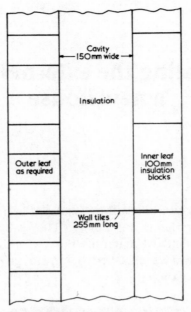

Diag. 37 **Insulated external wall—new** house

treated with water-repellent compounds so that if rain does penetrate the outer leaf it will not pass through the insulation.

Insulation materials which are blown or pumped into the cavity after the walls have been completed include mineral wool, sometimes called rockwool, expanded polystyrene beads and urea-formaldehyde-based foam. Your architect will advise you when making your choice on which insulation method and material to use in your particular house.

EXPOSED HOUSES

Problems can arise in some areas as, strictly speaking, the current Building Regulations call for a cavity in the wall, and filling the cavity, even with a water-repellent material, means that there will be no cavity left on completion. Until the regulations are altered to allow for this, most local councils will permit you to install cavity wall insulation. If, however, you are having your house built in a very exposed area, then they may not permit you to do this. Should this be the case, then an alternative wall construction is shown in diagram 38 and this includes both the cavity and the insulation thickness required. It will, of course, be more expensive than a standard wall construction, but will quickly pay for itself in terms of large fuel savings and additional comfort.

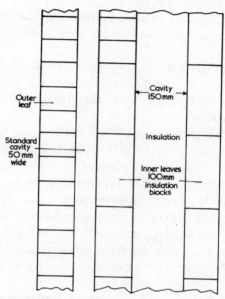

Diag. 38 Insulated external wall for a house in a very exposed area

General points

The thickness of the exterior walls has been increased slightly as a result of insulating them and to obtain the same interior dimensions the external lengths of the walls should be increased to compensate for this. Check the size of your rooms carefully on the building plan and ask the builder to show you a similar house with all the furniture in. Adding a metre to the length or width of the house will not materially increase the building cost and it can be used to turn impossibly small rooms into acceptable ones and acceptable rooms into comfortably sized ones. In particular, beware of tiny bedrooms, bathrooms and kitchens—they are just not acceptable by modern standards.

The minimum ceiling height permitted is 2300mm and sometimes this can give an out-of-balance look to a larger room or a through lounge and dining-room as well as restricting your choice of overhead lighting. Check that your architect is not working to minimum standards and that the ceiling heights are in proportion to the size of the rooms and the style of the house.

Keeping halls warm is always a problem. It helps considerably if there is an outer and an inner door and sometimes a porch may be needed to fit this arrangement in, but it is well worth while.

Ground floor partition walls are better built using concrete blocks 100mm, or even 150mm thick rather than from lightweight blocks. This gives a thermal storage bank which can give out heat on sudden cold days and absorb heat on hot days to give a much steadier temperature inside the house. An additional benefit is greater rigidity and better sound-proofing between the rooms. Upper floor walls built over the lower walls are better constructed from concrete blocks as well, but partition walls built over joists will have to be of lightweight materials.

A newly built house takes quite a lot of heat to warm it up and to dry it out. If you move in to it in the summer, this is no problem. If the house is completed during the winter the heating system, because it is of low output, will take quite a time to warm up the house. You should arrange for the heating system to be on for three or four weeks before moving in. All the internal doors should be left open to allow air to circulate and the night ventilators should be open to allow water vapour to escape.

14 Insulating the external walls of your present house

Before considering any insulation work you must first check all the exterior walls carefully for signs of defects. Usually this will be concerned with damp walls and you must check the following points before carrying on—remember, that insulating a wall will not cure damp caused by faults in the house and it may well make matters worse.

Are there any damp patches on the walls inside the house upstairs? Find out the cause and rectify the trouble. Common problems are blocked or leaking gutters or down pipes, loose or missing slates, loose or missing mastic work around the windows or, indeed, loose or missing mortar between the brickwork. Are there any damp patches on the walls inside the house downstairs? Again find out the cause and rectify the trouble. If the damp is above waist level then the causes could be similar to those just mentioned. If the damp is below waist level then it may mean that the damp-proof course in the brickwork, to stop damp rising from the ground, has failed or been bridged. Early houses have no damp-proof course and you may have to have one installed. Do not be confused with condensation which forms on the walls, particularly in kitchens and bathrooms, when a lot of steam has been created. The tendency for this to occur will almost be eliminated when the walls have been insulated.

Insulating houses with cavity walls

Timber joists shrink very slightly as they dry out in a warm house and this, together with the vibration caused by people walking around the house, loosens the mortar round the ends of the joists in the wall, and gaps between the joists and the brickwork result. Sometimes when plumbing alterations are done and a pipe runs through the wall, it is difficult to seal round the pipe properly where it passes through the inner leaf, perhaps because it is hidden under the floorboards or a bath. Inevitably some draughts blow up through the inside of the cavity wall into the spaces between the first floor joists and the only way to stop this is by cavity wall insulation.

As the cavity in the wall will only be about 50mm wide the insulation in the cavity by itself is not sufficient and additional insulation must be fitted to the inside of the exterior walls to bring the total thickness of the insulation up to 100mm. The work can be considered to be in two parts. Firstly, there is the cavity wall insulation and this has to be done professionally by a contractor. Secondly, the inside of the walls is to be panelled with additional insulation and this can be a do-it-yourself job, unless you wish to employ a tradesman to do it for you.

Insulating the cavity in the walls

This work must be entrusted to a recognised contractor. There will be a list of contractors in your local yellow pages under the heading 'Insulation'. Always choose a reputable company and they should be able to verify that their product is covered either by an *Agrément* certificate or a British Standard specification verifying the quality and suitability of the materials they use.

They will carry out the following procedure. First of all, holes will be drilled in the outer leaf of the walls at regular intervals and the insulating material will be blown or pumped into the cavity through the holes to fill it completely. The holes will be filled with cement or mortar and painted to match the brick or stone work. The men will also ensure that the air bricks underneath the ground floor are all clear and not blocked; some firms replace the existing air bricks with new sleeved ones. Usually the work is completed within a day for a smaller house and little mess results from the process.

The types of insulating materials generally used fall into the following categories: a water repellent mineral wool which is blown into the cavity; expanded polystyrene beads which are again blown into the cavity; or, a urea-formaldehyde-based foam which sets very like expanded polystyrene once it is in the wall.

PROCEDURE

You should obtain quotations from about four companies, say two offering foam insulation, one offering the bead insulation and one offering mineral wool (sometimes called rockwool) insulation. Compare them carefully and make your choice; sometimes discounts are given for installations carried out in the early summer. Ask for the name of a householder in your area who has had his walls insulated by the contractor and contact him, or her, to verify that the work was carried out to their satisfaction.

Very occasionally these companies will decline to insulate your walls either because the walls are in poor condition or because the house is on a very exposed site. If this happens you would be well advised to move to a more suitable house!

It is also best to check with your local council to see whether planning permission is required for cavity wall insulation in your area. Usually the insulation contractor is required to notify the council that they are going to install cavity wall insulation in your house and you should check this point with them.

Because this work can be done very quickly by the contractor immediate benefits will result and you should certainly have the cavity walls insulated before next winter. This will leave you free to concentrate initially on the roof and window insulation and on the draught prevention. The floor insulation and interior panelling can be done as each room becomes due for re-decorating.

Panelling cavity walls

The first-floor walls can be panelled at any time but on the ground floor the floor insulation must be done *before* the walls in these rooms are panelled. The panelling fits around the inside of the rooms and, of course, only the exterior walls need to be done.

Timber battens will be secured to the walls and the spaces between them filled in with insulation. A layer of polythene over this forms a vapour barrier to prevent water vapour from the room reaching the cold brickwork where it would otherwise condense behind the panelling. Supalux-type boarding is nailed over this to provide a hard durable surface suitable for decorating. A cross section through the wall is shown in diagram 39.

Because there is already 50mm of insulation in the cavity wall, the panelling insulation only needs to be 50mm thick to bring the total thickness of insulation up to 100mm. The only difference between panelling a cavity wall and a solid wall is in the thickness of the materials used. Follow the instructions for solid walls but use the reduced batten and insulation thickness as follows:

The battens should be 50mm × 50mm in place of 100mm × 50mm

The insulation should comprise two layers each 25mm thick in place of two layers each 50mm thick

The nails securing the battens to the wall will be 100mm long in place of 150mm long.

Panelling solid walls

Sometimes in houses with solid walls, heavy driving rain can soak the wall and penetrate right through, spoiling the plaster-

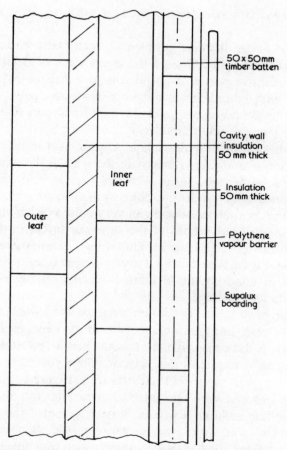

50 x 50mm
timber batten

Cavity wall
insulation
50 mm thick

Inner
leaf

Insulation
50mm thick

Outer
leaf

Polythene
vapour barrier

Supalux
boarding

Diag. 39 **Horizontal section through an insulated cavity wall**—existing
house

work and decorations on the inside. It may only occur in the
wall facing the prevailing wind. Assuming that none of the
faults listed previously are applicable, do you have this
problem? If the exterior walls are always dry then you have
no problems and the walls can be insulated as described
shortly. If you do have trouble with rain soaking through the
walls then this must be rectified *before* the insulation can be
installed.

You should first obtain advice from an architect or a firm

99

which specialises in these problems, on the best way to treat your particular walls. One of the latest ways of dealing with walls which give particular trouble is to panel the exterior of the wall with completeiy waterproof fibreglass panels. These can then be skimmed over and the final finish can be made to resemble stone, brick or rendering.

A further problem with walls is damage as a result of frost. This occurs when water remains in the wall in the winter and it can arise either as a result of water vapour from the warm house condensing inside the cold wall or from rain driving into slightly porous stonework. In very cold weather the water freezes and cracks sections of the stone, or brick, in the same way that frozen pipes burst. The stone becomes even more porous and is further damaged every time it freezes. In a wall which is in good condition there is insufficient water in the wall to do any damage.

If your external walls contain stone or brickwork which is obviously soft and crumbling you must seek professional advice as to the cause. It may be that just a few stones need replacing or it may be more serious. Once you have had the necessary repairs done and had the wall declared sound then you can proceed with the internal panelling and insulation. The panelling includes a vapour barrier so that if the cause of the problem was water vapour from inside the house, this will prevent any further water vapour entering the wall. On the other hand, if the trouble was caused by rain entering the wall, the remedial work should have included a waterproofing treatment to the outside of the wall to prevent the damage recurring.

DESCRIPTION OF THE PANELLING

Timber battens will be secured to the inside of the exterior walls and the spaces between them filled with insulation. A layer of polythene over this forms the vapour barrier to prevent water vapour from the room reaching the cold brickwork where it would otherwise condense behind the panelling.

Supalux boarding, or a similar type, is nailed to the battens to provide a hard durable surface suitable for decorating.

The floors of ground-floor rooms must be insulated before these rooms can be panelled. In the first-floor rooms there is, of course, no need for floor insulation and the panelling can be done at any time.

A cross-section through the wall and the arrangement of the battens are shown in diagrams 40 and 41. The construction avoids the need for plastering and gives an adequate finish for most purposes. However, if plastering is no problem to you, then plasterboards could be used instead of the Supalux boarding, but they will need to be skimmed over with plaster on completion.

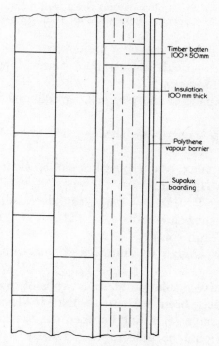

Diag. 40 Horizontal section through an insulated solid wall—existing house

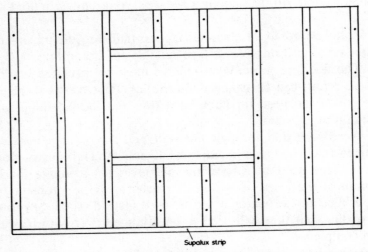

Supalux strip

Diag. 41 Layout of the timber battens for a wall 4.25m long and 2.6m high with a window opening 1.5m wide by 1.2m high—existing house

MATERIALS REQUIRED

Timber battens—rough sawn 100 × 50mm, pressure-treated against rot and insect attack

Insulation—expanded polystyrene, grade SD with FRA, 50mm thick

Vapour barrier—polythene sheeting 125 microns thick, 4m wide

Panelling—Supalux, Masterboard or similar board 9mm thick, sheets 2400 × 1200mm

Nails—150 and 100mm long—round head, 30mm-long galvanised plasterboard nails

Dowelling—12mm round

Miscellaneous—timber for finishing off the window surround and sill.

As an alternative insulation material, rolls of mineral or glass fibre which has been treated so that it does not absorb moisture are made for fitting between the battens of panelled walls. You should be able to obtain these from the larger builders' merchants in your area. The polythene sheeting can be obtained from a builders' merchant or some better DIY stores, and the timber from either a timber merchant or, again, some DIY stores. For the expanded polystyrene you will have to go to a board merchant—look up your local Yellow Pages under 'Boards'—and they will also supply the panelling board.

You should work out the quantities needed for your particular walls but, as a guide, the wall shown in diagram 41 which is 2.6m high and 4.25m long, with a window opening 1.2m high by 1.5m wide, would need the following quantities of materials—

Battens	25m
Polystyrene	6 sheets
Polythene	4m
Panelling board	3 sheets

PREPARATION

Roll back the carpet from the wall to be worked on, remove the curtains and rails and cover furniture with a protective sheet. Electrical fittings must only be removed and replaced by a competent person. Isolate the cables and remove any socket outlets in the skirting-board or wall. Pull the cables back below the floor or cut them off if they are embedded in the plaster. Similarly, isolate the cables and remove any wall lights or switches. Pull the cables back, above or below the floor as appropriate, or again, cut the cables if they are embedded.

Central heating radiators on a wall to be panelled will also have to be removed and refixed back onto the finished panelling: turn off the fuel and electric supplies to the boiler and turn off the cold-water feed to the separate header tank for the system; drain the central heating system; remove the radiator and alter the pipework to suit the new radiator position before doing the panelling. If there is more than one radiator in the room it should be borne in mind that this may very well suffice once the house has been insulated, allowing the other radiator to be dispensed with.

Remove the wooden architraves from around the window or external door if there is one in the wall. Prise them off carefully trying not to damage them as they will be refixed later. Prise off the skirting board in the same way. Once this has been eased away from the wall you may have to cut it in the middle to get it out. Try not to damage the board as it will also be fixed back onto the finished panelling.

Now check the condition of the wall plaster—if it is firm, leave it alone. However, if it is already loose and coming away from the wall, chip and scrape it off with a broad cold chisel. Fill any gaps or holes in the wall or brickwork, particularly where the architraves and skirting-board have been, with a cement mortar.

PANELLING INSTRUCTIONS

Chip away the plaster for a height of 15mm just above the floor boards, cut lengths of panelling board 115mm wide and nail them to the floor, butting them close up against the brickwork. Fill the gap between the board and the wall plaster with a plaster filler. This prevents draughts blowing up behind the panelling and also delays the breakthrough of fire in case of an underfloor fire.

Now cut the timber upright battens so that they are a snug fit between the boarding and the ceiling. Do not make them too tight for the ceiling plaster or the boarding could be damaged. Drill 5mm holes through the battens at about 600mm intervals to take the securing nails and knock a 150mm nail into each hole. Place the batten against the wall in the desired place and lightly tap the nails into the wall to mark the positions for the drilled holes for the wooden plugs.

Drill 12mm holes into the brickwork for a depth of 75mm. It is well worth hiring an electric drill with a hammer action if you do not have one and you will, of course, need a masonry drill bit. Clear out each hole, drive in a piece of 12mm dowelling and saw the end off flush with the wall. Replace the battens in their correct positions and drive the nails into the plugs in the wall. The nail heads should be just below, or level with, the surface of the timber.

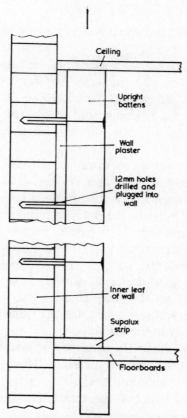

Diag. 42 Method of fixing the battens to the wall—existing house

To make the most economical use of the panelling board it is sometimes better to draw the battens and the boards on a piece of paper first to see how you are going to cut them. The battens should be spaced not more than 600mm apart and joints between each board should be over the centre of a batten so that the edges can be securely nailed. The nails securing the battens to the wall should be about 200mm down from the ceiling, and up from the floor, with the intermediate ones being about 600mm apart. Cut across battens to fit along the underside of the window sill and above the window. Nail them to the upright battens on either side. Now cut vertical battens to fit below and above the window, if they are required, and make sure that the cross battens are straight and level. Drill and plug the vertical battens to the wall and nail the cross battens to them. Be careful when drilling the wall around the window—angle the hole into the brickwork or you may find that you are just drilling into the plaster surrounding the window.

Any electrical wiring required should now be installed and socket outlets should be in metal boxes. It is also better if the cables are run in metal conduit which is correctly fixed to the boxes. Seal any holes in the floor or ceiling where the conduit or cables pass through them with a plaster filler.

Now cut the expanded polystyrene neatly with a hacksaw blade or a sheet saw so that it is a snug fit between the battens—try to keep the edges straight. Two layers of insulation are needed so that any gaps in the first layer will be covered by the second layer. Butt all joints snugly together and stagger joints in the two layers.

Tack the polythene sheeting to the battens, to cover the whole wall in one piece if possible. If a joint is necessary overlap the polythene and seal the joint with polythene tape available from a builders' merchant. The sheet will cover the window opening—do not cut this out yet. This sheet forms the vapour barrier to prevent water vapour from the room reaching the wall. Make sure the sheet overlaps all the adjoining walls, ceiling and floor by at least 25mm.

Cut and nail the panelling board to the battens. All joints must be over the centre of a batten so that the board edges can be securely nailed. If you have particularly high ceilings you may have to put in some extra cross battens. Nail every 150mm and use a punch to drive the nail head flush with the surface of the board. If you try to do this with a hammer alone you are sure to damage the board. Butt all the edges neatly together.

Now cut out the polythene sheet from the window opening by cutting it diagonally from the centre to each corner. Fold each piece back onto the window. Trim the sides and top of the window reveal with plywood, panelling board or other timber and trap the polythene beneath this to continue the vapour barrier. For the window-sill you could use either of the constructions shown in diagrams 43 and 44.

Diag. 43 Possible window-sill construction

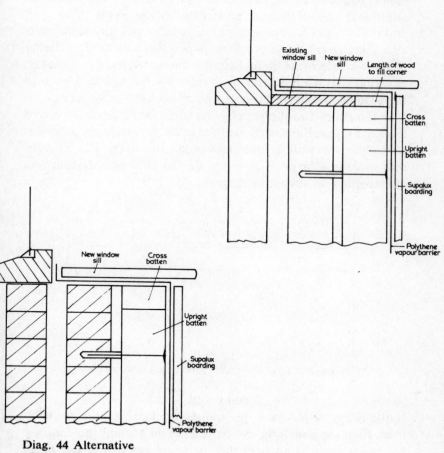

Diag. 44 Alternative window-sill construction

Cut off the excess polythene all round with a sharp knife and fill all gaps and joints with a plaster filler. Refix the ornamental window surrounds and the skirting-boards by nailing them to the timber battens through the boarding. Fill in any holes in and around the timber work.

The panelling is now complete. Spot paint the nail-heads with an oil-based primer and paint the whole with an oil-based sealer, followed by an undercoat and gloss coat; this forms an additional vapour barrier and seals all the joints. You may find that the finish obtained is quite satisfactory, but if the nails and joints tend to show, then paper the wall with a lining paper and decorate it to suit the room. Refix the radiator, ensuring that all the securing screws for the brackets go into the battens. Similarly connect up any sockets or switches.

If the room has more than one outer wall, continue round the room panelling each wall in exactly the same manner. Complete one wall before starting on the next. This ensures that there is a layer of panelling board between adjacent sets of panelling as shown in diagram 45.

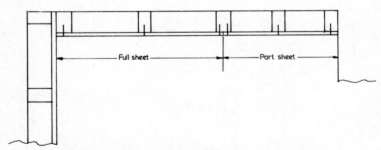

Diag. 45 Layout of the panelling in an alcove

CHIMNEY BREASTS

If the fireplace has been previously bricked up and plastered over, then the panelling can be carried on around the chimney breast. If there is an open fire place or gas fire in use panel up to the chimney breast but *not* around it.

OLDER HOUSES

If there are ornamental mouldings around the top of the room where the walls meet the ceiling, cut the vertical battens until they blend in with it. On completion, some or all of the moulding will be hidden behind the panelling. Nothing much can be done about this, but by the time the decorating has been completed the slightly odd look is usually acceptable. A simple cornice can obviously be replaced with a new one.

NOTES

Bathrooms and toilets will frequently present a problem because the bath and toilet have been fixed adjacent to an outside wall. To move these may mean plumbing alterations and the same applies to kitchens where the sink is on an outside wall.

However bathrooms and kitchens are the rooms where most condensation occurs on the walls and, from this point of view, they are the ones which will benefit most as a result of correct insulation. The interior wall surface of an insulated wall is at room temperature and it is unlikely that condensation will form on it. If the suites or sink are likely to be replaced in the future, then the panelling should be done at the same time. If moving these items presents no problem to you then, of course, the work can proceed all the sooner.

Kitchen wall units can be very heavy when they are full of crockery and they should be screwed right through the panelling board securely into the timber battens behind. If it is known that wall units are to be fitted then, while the battens are being fixed to the wall, additional cross battens can be included, if necessary, to coincide with the fixing holes in the units. A batten must be fixed to the ceiling joists to prevent any possibility of the panelling pulling away from the wall. This is shown in diagram 46.

Wall panelling sheets based on hardboard or plywood are available with a range of attractive finishes and they are quite popular. If you wish to finish your walls with these sheets then

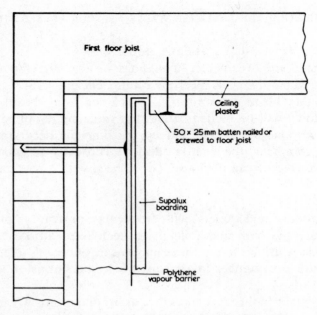

First floor joist

Ceiling
plaster

50 x 25 mm batten nailed or
screwed to floor joist

Supalux
boarding

Polythene
vapour barrier

Diag. 46 The batten fixed to the joists, if kitchen wall units are to be
fixed to the upright panelling battens

they must be secured *over* the panelling boards. The panelling
boards must not be omitted as the thin decorative sheets have
little resistance to fire.

15 Heat losses through the ground floor

Basically there are only two types of ground floors in houses. Firstly, there are solid floors, which are concrete in newer houses, or stone slabs in very old houses. Concrete is laid on a prepared base of broken stone, but old stone floors are laid straight on to the earth. Secondly, there are timber-suspended floors where timber joists are supported clear of the ground by brickwork. The joists are then covered with floorboards.

Both concrete and stone conduct heat to the ground below. Timber, when thick enough, is a reasonably good insulating material. However, the floorboards are relatively thin and allow heat from the room to pass through. There is a flow of air under the floor to keep it dry and this cools the underside of the floor ensuring a rapid loss of heat. In addition, timber floors sag slightly with age and use. They expand and contract depending on how much moisture there is in the air in the house and also with changing temperature and, inevitably, small gaps occur around the edges of the floor allowing draughts to blow through them. In both types of floor it is necessary to install a layer of insulating material to reduce the heat loss to an acceptable level.

Concrete floors laid in the latter half of this century will already have a damp-proof layer built into them and they should be quite dry. Older floors and particularly stone floors will not have this and may well be damp. For these a damp-proof layer will be included in the construction to bring the floor up to modern standards.

Heat loss figures

The heat loss figures for both solid and timber floors is about 0.0007 and this compares with the maximum figure of 0.001 which is allowed by the 1976 Building Regulations for new houses. Adding insulation reduces this figure substantially and the effect of various thicknesses of insulation is shown below:

TABLE 18
EFFECT OF VARIOUS INSULATION THICKNESSES
ON GROUND FLOORS

Insulation thickness mm	Heat loss figure for solid and timber floors	Reduction %
None	0·0007	—
25	0·00046	35
50	0·00035	50
75	0·00028	60
100	0·00023	66

Because heat rises then, given similar conditions, the amount of heat lost through a ground floor will be less than that lost through the roof. This partly accounts for the difference in the heat loss figures for completely un-insulated floors and roofs which are 0.0007 and 0.0019 respectively. The thickness of insulation in a floor need not therefore be quite so high as that in a roof and in both new and older houses 100mm of insulation will give good results.

Heat loss calculations

The ground-floor rooms should be at 20°C for comfort and this figure is used in our comparisons in table 19 on the opposite page.

By comparing these figures you see that the amount of heat lost has been reduced by the insulation to one third that of a normal floor design.

TABLE 19
AMOUNT OF HEAT LOST THROUGH THE GROUND FLOOR WHEN IT IS FREEZING OUTSIDE

	Typical house size			
	2-bed-roomed terraced	*3-bed-roomed semi*	*4-bed-roomed detached*	*5-bed-roomed detached*
Total floor area sq m	30	40	65	90
Amount of heat lost through an un-insulated floor				
Per hour kW	0·4	0·55	0·9	1·2 ($\times$ 0·0007 $\times$ 20)
Per day kW hrs	6	9	14	19 ($\times$ 16)
Reduced heat loss through an insulated floor				
Per hour kW	0·15	0·2	0·3	0·4 ($\times$ 0·00023 $\times$ 20)
Per day kW hrs	2	3	5	6 ($\times$ 16)

16 Insulating the ground floor of a new house

Timber-suspended floor

As shown in diagram 47, expanded polystyrene sheet will be installed between the ground-floor joists to the correct thickness. It is not necessary to include a vapour barrier because any water vapour passing through the floor will be dissipated by the air currents under the floor.

INSULATION SPECIFICATION
Expanded polystyrene, grade SD with fire retarding additive (FRA), thickness 50mm.

INSTALLATION PROCEDURE
Naturally, this will be done by your builder to the architect's

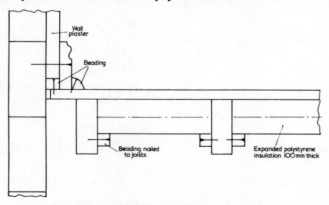

Diag. 47 Insulated timber-suspended floor

114

instructions and the next few paragraphs are a guide as to what is required in the way of fitting the insulation. The ground-floor joists will have been installed as the building work is done and the insulation should be fitted at the same time as the floorboards are being laid.

Battens are nailed along the full length of each joist, 98mm down from the top of the joists to ensure that the insulation is just nipped when the floorboards are laid. The polystyrene should be cut to be a snug fit between the joists and two layers are needed to give the correct thickness of insulation. Joints must be closely butted together and staggered between the two layers and the insulation should be butted up snugly to the brickwork at the end of each run. Fitting the insulation in two layers means that any small gaps in the first layer will be covered up by the second layer.

Insulate part of the floor at a time and lay the floorboarding over that area before proceeding, so as not to damage the insulation. On two opposite sides of the floor there will be joists running close to the walls and it will not usually be possible to fit insulation between these and the adjacent wall.

All timber floors sag with age and loading and, to prevent draughts blowing up around the edges of the floor, square beading, of section to suit the plaster depth, should be nailed to the floor boards against each wall as in diagram 47. The seal is completed with quarter-round beading, again nailed to the floor only, fitted firmly up against the skirting-board.

ALTERNATIVE INSULATION MATERIALS

It is possible to insulate between the joists with materials such as fibreglass or mineral fibre but asbestolux-type boarding will have to be cut and fitted in between the joists, resting on the battens, to retain the insulation.

Solid concrete ground floor

A section through the finished insulated floor is shown in diagram 48 and a normal concrete base is first laid incorporating the damp-proof layer. When all the heavy building work has been completed a secondary floor is laid incorporating the insulation and a vapour barrier.

MATERIAL SPECIFICATIONS
Insulation—expanded polystyrene, grade SD with FRA, sheet thickness 100mm (or two layers each 50mm thick)
Vapour barrier—polythene sheet 250 microns thick.

INSTALLATION PROCEDURE
Again this will be done by the builder to your architect's instructions and the following is only a guide as to the procedure required. The levels of the foundations, damp-proof course and base concrete will be worked out by your architect to suit the particular site, bearing in mind the extra thickness of the finished insulated floor. A section through the floor is shown in diagram 48.

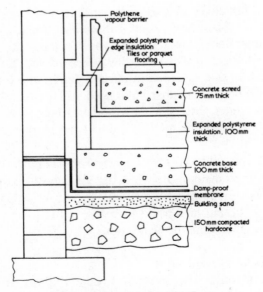

Diag. 48 Insulated solid floor

The first stage is to install the base concrete and this is really just a standard concrete floor. The order of laying it comprises 150mm of compacted hardcore covered with building sand spread evenly and level, a damp-proof membrane lapped into the walls with the damp proof course and finally 100mm of concrete, trowelled level and smooth.

The second or next stage should not be done until the main constructional work and the roof have been completed. The finished insulated floor could be damaged if heavy stones, blocks, beams or scaffolding were accidently dropped on to it from a height.

Cut the edge insulation 175mm deep from 25mm-thick polystyrene and place it round all the walls. Lay 100mm of insulation over the whole floor area fitting it snugly to retain the edge insulation in place. Cover the whole area with one layer of 250 microns thick polythene sheet, overlapped up the walls by 200mm to form a vapour barrier, and seal any joints with polythene tape. Finally, lay a 75mm-thick concrete screed trowelled level and smooth. In practice, of course, the insulation and screed have to be laid in sections otherwise there would be nowhere for the men laying the floor to stand.

You will now have a high quality, damp-proof and insulated floor, over which you can lay virtually any floor-covering you like. Carpets and vinyl should have underlays but tiles can be cemented straight to the floor. You could even have a timber or parquet floor as shown in the diagram.

17 Insulating the ground floor of your present house

Timber-suspended floor

As in a new house, the insulation will be fitted between the joists to the correct thickness. Some floors have easy access beneath them and others have no access beneath. Instructions are given for both situations in this chapter. Diagram 49 shows the insulation in position between the joists.

MATERIALS REQUIRED

Insulation material—expanded polystyrene, grade SD with fire retarding additive (FRA), thickness 50mm
Wood beading—sawn 25mm square
Nails—50mm long.
Again, the expanded polystyrene can be obtained in sheets from a board merchant.

Insulating a timber floor with access underneath

PREPARATION

As with the roof this part of the house is looked at very infrequently so now is the time to carry out some checks to make sure that all is well with the ground floor. The first of these is to roll the carpets back a short way from each wall in turn and check whether there are any signs of woodworm in the floorboards. From below again, check for any signs of woodworm attack in the joists. Look out for rotting timbers—

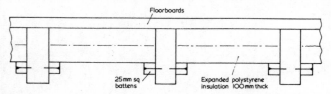

Diag. 49 Insulation fitted between the joists of a timber floor

118

particularly the ends of the joists where they go into the walls. Try to push a medium screwdriver into the joists at intervals —if it only makes a small dent then the timber is probably sound, but if it goes in easily for some way you must seek professional advice.

Make sure that all the ventilation bricks are clear. An unobstructed flow of fresh air under the floor is essential to prevent damp and subsequent rotting of the timbers. Check also to see that the wiring is the modern type with a grey plastic finish. If it is any other type, have it checked by an electrician because it is time that it was replaced.

Use a powerful hand lamp or a car-type inspection lamp. With the latter, ensure the cable is in good condition and try not to crawl or kneel on the cable. Do not go under the floor if you are likely to be taken ill by reason of age or other cause.

From outside, there should be no soil above the damp-proof course in the brickwork. If there is, clear it away to at least one whole course below the damp-proof course.

INSTALLING THE INSULATION

Working from below the floor, cut the polystyrene with a hacksaw blade or sheet saw, to be a snug fit between the joists. Because it is rarely possible to take these sheets beneath the floor this is best done as a two-man job. One will be below the floor measuring the spaces between the joists and fixing the insulation; the other will be above the floor cutting the insulation carefully to the size required and passing the strips down.

Fit the insulation between each pair of joists, two layers thick, to make a total thickness of 100mm. It should be a snug fit both between the joists and against the walls at the end of each run. Press the insulation firmly up against the floorboards and secure it in position with the square beading nailed to the joists along both sides.

Continue filling in all the spaces between the joists, butting joints in the insulation snugly together and stagger the joints in the two layers. Measure each pair of joists separately as the spaces between them will vary. On two sides there will be joists running close to and parallel with the walls and it will not usually be possible to fit insulation in these two spaces.

Insulating a timber floor with no access underneath

There are theoretically two possible ways to insulate these: either to lay an insulating layer over the existing boards and reboard over the insulation, or to lift the floorboards and fit the insulation between the joists as for a new house. Whilst the former may initially sound simpler it has the disadvantage that it reduces the ceiling height and all the doors will consequently have to be altered. In addition, new floorboarding will have to be bought when you already have a perfectly good set. Lifting the floorboards may sound unattractive, but it is not, in practice, difficult to do. In addition, the minimum amount of new materials have to be purchased and this is therefore the method which will be used. It would be sensible to do this work when the room is due for re-decoration.

LIFTING THE FLOOR AND CARRYING OUT THE CHECKS

Clear the room of all furnishings and isolate the cables to any socket outlets in the skirting-boards; remove the socket outlets. Again, electrical work must only be undertaken by a competent person. Now prise away the skirting-boards from the walls, trying not to damage them as they will be fixed back later.

Cut the tongues on both sides of a floorboard running across the centre of the room. This can be done quickly and neatly with either an electric circular saw set to a depth of 15mm —do not set the saw any deeper or you may cut into electric cables or the joists—or with a tongue-cutting tool, which you should be able to hire from a good tool-hire firm.

Now switch off the electricity and prise up the board. Number all the boards in one half of the room so that they can be re-laid in their original order. Prise up the boards in one half of the room only. Take care not to damage any electric cables, gas or water pipes—you should feel under each board before lifting it. A floorboard-lifting tool—rather like a large crowbar with a pronged end—will make light of this work, if you can hire one. If you are careful you may be able to lift the remaining boards without cutting all the tongues, but if you find that the boards are splitting, then it would be sensible to cut the tongues first. Take out any nails left in the boards or joists.

Check all the timbers for any signs of woodworm attack or rotting, particularly the joists where they go into the wall. Check that the wiring is the latest type with a grey plastic finish—if it is any other type, have it checked by an electrician, because now is the time to replace it. It is also well worth while at this stage to move any socket outlets to a more convenient height up the wall rather than to replace them in the skirting-boards later.

Check to see that all the central heating pipes and water pipes under the floor are properly lagged and that there are no leaks.

INSTALLING THE INSULATION AND RE-LAYING THE FLOOR

First nail 25mm-square beading 98mm down from the top of the joists to retain the insulation and ensure that it will be just nipped when the boards are replaced. Cut the polystyrene to be a snug fit between the joists; two layers are needed to bring the total thickness of the insulation up to 100mm. Joints between the two layers should be staggered and all the edges should be butted snugly together and up against the brickwork at the end of each run. On two sides of the room there will be joists running close to and parallel with the walls and you will not be able to fit the insulation between these joists and the wall.

Place the floorboard nearest the wall back in position and

nail it down. Replace the remaining boards in the correct order and nail them down making sure they are pressed hard up against each other. Take care not to nail through any cables or pipes. Take care also that you do not stand on the polystyrene—it will give way. Stand only on the joists; preferably lay a few loose boards across the joists to stand on whilst you are working.

Now lift and relay the second half of the floor. The boards in this half should, of course, be replaced by pressing them hard against the boards already laid. If you have had to cut the tongues you will probably find that on reaching the wall there is a gap—fill it in with a new board sawn to fit.

Nail square beading, of the same size as the plaster-depth, all round the edge of the floor, as shown in diagram 47, and fill in any gaps in the plasterwork with a cement mortar. Replace the skirting-boards, as described on page 127 and nail quarter-round beading, pressed firmly against the skirting-board, to the floor only to complete the draught sealing.

Finally, replace any electric sockets which have not been dealt with.

Solid floors

If the floor is in good condition and fairly level, both the walls and floor are normally dry and the ceiling height is at least 2400mm, then the floor is ideal for insulating. A secondary floor will be laid on top of the existing floor and it will include a damp-proof layer and vapour barrier as well as the insulation and a new, hard floor-surface.

If, however, the floor is rough, already breaking up and damp, or the ceiling height is under 2400mm then, obviously, the floor is already a problem. It can only be resolved by breaking up the floor and re-laying it to modern standards before the insulated floor can be built. (Outline instructions are included for this work but you should seek further advice if you have not mixed and laid concrete before.)

If the walls are wet or damp at the bottom, there would

appear to be either no damp-proof course or else it has been bridged or broken. The cause must be resolved by seeking professional advice, as insulating the floor will only aggravate the problem.

An obvious point to check first is that there is no build-up of soil against the outer wall. Clear away any soil to at least one whole course below the damp-proof course. If there is no damp-proof course, then clear away the soil for at least two or three courses below the level of the floor—if this is possible. Give the wall a few weeks to see if it will dry out sufficiently to remove the damp from the room.

If the dampness is in a cellar wall you may have to dig out all the soil right down to the level of the foundations and install a waterproof barrier against the wall before replacing the soil.

SOLID FLOOR IN GOOD CONDITION

An additional floor will be laid on top of the existing concrete and a choice will have to be made now on the type of finished floor-surface required. There are two possibilities: either a timber floor—this is the easier to lay and flooring grade chipboard panels are used to give a durable finish—or a concrete floor. Concrete is better if the floor is to be tiled later as the more rigid the surface is on which the tiles are laid, the better. Care must be taken laying the concrete and a certain amount of expertise is required to obtain a smooth and level surface.

Both constructions include a damp-proof layer, an insulation layer, a vapour barrier and the floor-surface material. A vapour barrier is again necessary to prevent water vapour from the warm room penetrating through to the cold original concrete where it would otherwise condense and a build-up of water could occur under the insulation.

Both constructions raise the level of the floor—the timber floor raises the level by 125mm and the concrete finish raises the level by 165mm. The doors will now have to be checked as, clearly, raising the floor reduces the door height and it is not usually desirable to reduce the height of the doorways below 1800mm. The minimum increase in floor height is given with the timber finish. If the present doorway height is 1950mm or more, the floor level can be raised, bearing in mind that on completion the new reduced doorway height should be acceptable for most people. If it is 1875mm it may well be worth sacrificing a small amount of insulation by reducing the insulation thickness to 75mm, to avoid digging up the floor. The doorway height would then finish at 1775mm. If it is 1800mm any increase in floor height will result in an unacceptably low doorway—unless the room is for use as a children's playroom, in which case a low doorway may have its attractions.

To remedy the last situation, two constructions are possible but both increase the extent of the work considerably: the floor can be dug up and re-laid at a lower level as described later, or, if the ceiling is high, there will be a substantial amount of brickwork above the doors and it is possible that the doors and frames can be removed and refixed after the lintel has been refixed higher up in the brickwork. Professional advice should be sought before doing this as there may be other supporting joists resting over the lintel.

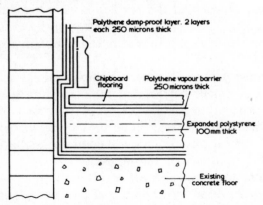

Diag. 50 Insulated chipboard floor laid on good existing concrete

Having decided that the existing concrete floor is in good condition and that any problems with the doors have been resolved, then you can proceed to lay an insulated floor on top of the existing concrete, with either a timber finish or a concrete finish as described next.

Insulated floor with timber surface
The construction will be seen from diagram 50 and it is fairly straightforward: a damp-proof layer is put down, followed by the insulation, a vapour barrier and the timber boarding.

MATERIALS REQUIRED
Insulation—expanded polystyrene, grade SD with fire retarding additive (FRA), 100mm thick (or two layers each 50mm thick), sheets 2400 × 1200mm
Damp-proof layer and vapour barrier—Polythene sheet 250 microns thick, sheet width 4 metres
Timber floor—flooring-grade chipboard, 22mm thick, sheets 2400 × 600mm or 2400 × 1200mm
Glue—woodworking adhesive

Again the polythene sheeting can be obtained from builders' merchants and the expanded polystyrene and flooring grade chipboard, from a timber and board merchant. Chipboard for use on a floor is tongued and grooved around the edges so that the sheets interlock when they are laid. As a guide, a floor 5m by 3½m would need the following quantities of materials:

Expanded polystyrene	6 sheets (100mm thick)
Polythene sheet	16m
Chipboard flooring	12 sheets (2400 × 600mm)

PREPARATION

Any socket outlets in the skirting-board must be removed and refixed higher up the wall and this should, of course, only be done by a competent person. If on an outer wall they must be refixed into the wall panelling as in chapter 14.

Remove wooden skirting-boards by prising them away from the walls—try not to damage them because they can be re-used. Similarly remove the wooden architraves around the doors by prising them away. If the door opens inwards it will have to be unscrewed from the frame.

Chip the plaster off the wall for a height of 150mm from the floor and fill any holes in the brickwork with a cement mortar mix. The floor should now be scraped lightly with a broad cold chisel, just to remove any sharp high spots which may cut the polythene damp-proof layer, and then swept clean.

LAYING THE INSULATION

Change into soft-soled shoes so as not to damage the polythene sheeting when you walk on it. Open out the polythene and lay one complete layer over the whole floor area. Overlap it up the walls by 200mm and fold it neatly in the corners—do *not* cut it. If possible, cover the whole floor area with one sheet; if a joint is necessary, overlap the polythene and seal the joint with polythene tape available from your local builders' merchant. Then lay a complete second layer of polythene over the whole floor staggering the joint if you have had to make one. This completes the damp-proof layer.

Put a complete layer of expanded polystyrene sheets over the whole floor, butting the sheets snugly together to make the whole a snug fit between the walls. Cut the sheets carefully with a hacksaw blade or sheet saw. As the sheets are being laid, place a few chipboard sheets on them to stand on. Do not stand directly on the polystyrene or it will be damaged. If you are using 50mm-thick polystyrene then a second layer will be needed to bring the total thickness up to 100mm; any joints between the two layers should, of course, be staggered.

Lay a sheet of polythene over the expanded polystyrene, again covering the whole area with one sheet, if possible; if a joint is needed, seal it with polythene tape. Overlap the polythene up the wall sides by 100mm and fold it neatly in the corners. This layer forms the vapour barrier.

Lay the first sheet of flooring-grade chipboard, starting from one corner, leaving a gap of 10mm between the board and the two wall sides. Lay further sheets, cutting them as necessary, to cover the whole floor area and glue all the joints as you proceed. Remember to leave a 10mm gap all round between the boards and the walls. This is necessary because the boards expand with changes in humidity and temperature.

Refit the door surrounds, after cutting them to suit the new floor level, and refit the skirting-boards, trapping any excess polythene behind them. To fix the skirting-boards, place them against the wall and drive 75mm nails into them at, say, 1000mm intervals, so that the nails just mark the wall. Remove the board and drill 9mm-diameter holes by 75mm deep, with a masonry drill, into the wall. Plug the holes with 9mm-diameter dowelling, replace the boards and drive the nails home. If necessary, cut the door to suit the new floor level and rehang it back in the doorway.

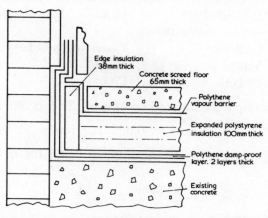

Diag. 51 Insulated concrete floor laid on good existing concrete

Insulated floor with concrete surface

The construction of this is shown in diagram 51 and, again, the arrangement is fairly straightforward. A damp-proof layer is put down, followed by the insulation, a vapour barrier and the concrete screed floor surface.

MATERIALS REQUIRED

Insulation—expanded polystyrene, grade SD with FRA, 100mm thick (or two layers each 50mm thick), sheets 2400 × 1200mm

Damp-proof layer and vapour barrier—polythene sheeting 250 microns thick, sheet width 4m

Concrete for the floor, mixed (by volume)—one part cement, two parts sharp (river) sand, four parts 10mm washed gravel.

The polythene sheeting and concrete materials can be obtained from your local builders' merchant, or some DIY stores, and the expanded polystyrene from a board merchant, as before.

As a guide, the following quantities of materials would be needed for a floor 5m by $3\frac{1}{2}$m:

Expanded polystyrene	6 sheets (100mm thick)
Polythene sheet	16m
Concrete materials—	
Cement	7 bags (each 50kg)
Sand	850kg
Gravel	1700kg

In this example the finished volume of the concrete is:

$$5 \times 3\tfrac{1}{2} \times 0{\cdot}065 = 1{\cdot}1\text{cu m}$$

The weights of materials required to make 1cu m of finished concrete are:

Cement	300kg
Sand	750kg
Gravel	1500kg

You should always round up the quantities needed, as has been done in the example above, to ensure that you have sufficient, and to allow for wastage.

PREPARATION

Any socket outlets within 300mm of the present floor must be refixed higher up the wall and, again, electrical work should only be carried out by a competent person. If they are on an outer wall then they can be fixed into the wall panelling later.

Remove wooden skirting-boards by prising them away from the walls, trying not to damage them because they will be re-used. Remove the wooden architraves around the door by prising them away. Also remove the door if it opens inwards onto the floor being worked on.

Chip the plaster off the walls for a height of 200mm from the floor and fill any holes with a cement mortar mix. The floor must again be scraped lightly with a cold chisel to remove any sharp high spots and then be swept clean.

LAYING THE CONCRETE FLOOR

Change into soft-soled shoes, open out the polythene and lay one complete layer over the whole floor area, overlapping it up the wall sides by 250mm. Fold it neatly in the corners. If a joint is necessary, seal it with polythene tape. Then, lay a second complete layer of polythene over the whole floor area, staggering the joint if you have to make one. This completes the damp-proof layer.

To prevent heat being lost from the new concrete floor surface to the ground, via the wall, a layer of insulation is needed around the edges of the concrete as well. Place lengths of 38mm-thick by 165mm-wide polystyrene along the wall sides and hold it in place with a complete layer of expanded polystyrene sheets fitted over the whole floor area. Butt the sheets snugly together to make the whole a snug fit between the walls. Cut the sheets carefully and, as the sheets are being laid, place a few boards on them to stand on. You must not stand directly on the polystyrene insulation or it will be damaged. If you are using 50mm-thick insulation, then two layers will be needed and joins between the layers must be staggered.

Lay a sheet of polythene over the expanded polystyrene, covering the whole area, and seal any joins you have had to make with polythene tape. The polythene should be overlapped up the sides of the walls 150mm and folded neatly in the corners to form the vapour barrier. Place the board you have been using to stand on, on top of the polythene as you go.

The concrete for the floor is going to be 65mm thick and, to obtain a level surface, it must be laid between two battens of this thickness, spaced about 1200 to 1800mm apart. The concrete is laid in strips, allowing each one to set before starting on the next one. You must not attempt to lay all the concrete in one piece, partly because it is very difficult to do and partly because you will inevitably stand on the insulation and damage it.

Place a 65mm-deep batten along one wall edge and a further batten about 900mm from it, placing bricks behind this latter batten to hold it in place. Mix and lay the concrete between the two battens, level it off, skim it smooth and leave it for twenty-four hours.

Remove the battens and continue concreting the floor in strips until the whole floor is covered. Stagger the joints in the concrete with those in the insulation and remember to keep boards over the insulation for walking on. Finally, fill in around the edges, refix the skirting-boards, door surrounds and door as previously described on page 127.

Solid or stone floor in poor condition
There is no option here but to dig up the floor and re-lay it correctly to modern standards. Obviously this involves more work but it is worth doing and the procedure is straight forward. The work is really in two parts. First the floor must be dug up and a solid base concrete floor re-laid at a lower level than that of the existing floor, incorporating the damp-proof

layer. Secondly, an insulated floor is laid on top of this with either a wood or concrete finish, almost exactly as described in the previous pages.

If the conditions are right, the finished level of the new insulated floor will be same as the old floor, so avoiding any problems with the doors or ceiling height. It might even be possible to finish up with an increased ceiling height if the ceiling is too low at the moment, but this can only be decided as the work proceeds—as explained later. If, however, the foundation for the new base floor cannot be dug sufficiently low, then the finished floor will be above the existing level. A section through the completed floor is shown in diagram 52.

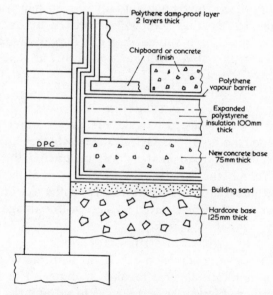

Diag. 52 Complete new insulated floor and base

MATERIALS REQUIRED FOR THE NEW BASE

Damp-proof layer—polythene sheet 250 microns thick
Stone hardcore—50mm crusher run

Concrete, mixed (by volume)—one part cement, two parts sharp (river) sand, four parts 20mm washed gravel
Building sand

As a guide, the following quantities of materials would be needed for a floor 5m by 3½m:

Polythene	11m
Crushed stone	3000kg
Building sand	1000kg
Cement	400kg (8 bags)
Sharp (river) sand	1000kg
20mm gravel	2000kg

The quantity of crushed stone actually needed may vary from this, because the existing floor will be broken up and used as hardcore. It has been assumed that this floor is 50mm thick; if it is thicker, less stone will be needed and if it is thinner, then more will be needed.

PREPARATION

Check that the main electric cable and gas or water pipes do not run under the floor. Also check to see whether any electric cables have been buried in the floor, by tracing the ends. If in doubt, switch the power off at the mains whilst you are working. Remove any socket outlets from the skirting-boards and remove the skirting-boards, trying not to damage them.

DIGGING UP THE OLD FLOOR

If the old floor is concrete, break it up with either a sledge-hammer or a hired, electric breaker hammer. The bit on a power hammer should not be allowed to go any deeper once it has broken through the concrete, as there is always the possibility that a drain may have been laid under the house. If the floor consists of stone slabs, they can be lifted with a crowbar and pickaxe.

Dig away and remove the subsoil underneath, for a depth

of 390mm below the finished level of the new insulated floor if it is to have a concrete finish, or 350mm if it is to have a timber finish. Do *not*, however, dig below the top of the wall foundations.

This is the limiting factor which governs the level of the finished floor: if the foundation level is not reached, then there are no problems and the finished floor will be in the desired position; if, however, the foundation level is reached before the required amount has been dug out, then the completed floor will finish above the old floor level—by the amount that the depth is short. A point to bear in mind is that, if the ceiling height was originally a bit low and the foundation level has not been reached, another 100mm could be dug out to increase the ceiling height slightly. The finished floor must not be below the level of the damp course in the walls if there is one.

Do half the floor at a time and check the levels carefully as you work. Throw in the broken concrete or stone slabs and break them up well. Top up with the hardcore to give a thickness of 125mm, well packed in. Again, check the levels carefully. Cover the hardcore with building sand, well packed in, to make sure that no stones project through the sand, for these would cut the polythene, damp-proof layer.

Change into soft-soled shoes and lay one layer of polythene over the whole floor area, overlapping it 300mm up the wall sides. Again seal any joints and fold the sheet neatly in corners to make it fit. It is essential that you do not cut it to fit the corners, or damp and water could work their way into the floor. Lay a complete second layer of polythene sheeting, staggering the joint if you have had to make one. This completes the damp-proof layer.

Mix and lay 75mm of concrete over the entire floor area and level it off carefully leaving a smooth flat surface. This has now completed the first stage of the work and the concrete must be allowed to dry out thoroughly before the next stage is undertaken. Concrete takes a surprisingly long time to dry right through and it should be left for about one month in the summer or two months in the winter.

INSULATED FLOOR SURFACE

A decision now has to be taken on the type of floor surface required and, again, there are two choices—timber or concrete. When you have decided which one to have, proceed exactly as described in pages 124 to 127 for the timber finish, or pages 128 to 130 for the concrete finish. The two layers of polythene sheet mentioned on those pages, which were included under the insulation as a damp-proof course, should be omitted, as they have already been included with the base concrete. The layer of polythene above the insulation must still be included, as this forms the vapour barrier to prevent condensation under the insulation.

NOTES ON EXPANDED POLYSTYRENE

On first sight it may strike you that this is a rather fragile material to use under a floor in which all the loading has to be carried by the insulation. It is, however, all a question of spreading the load. Standard grade (SD) expanded polystyrene will carry a working load of 3lbs per square inch. Obviously, if you stand directly on the polystyrene, then it will be damaged, but both the chipboard flooring and the surface concrete spread the load so that, with the chipboard, it will carry something like 1cwt over an area 6 inches square and with a concrete surface, which spreads the load even better, it will carry 4cwt over a square foot.

The chipboard spreads the load quite adequately for normal use, but if you are proposing to do any heavy work in the room, or if you are going to have a piano in the room, then it would be sensible to use a concrete finish. A chipboard floor can then be laid on top of the concrete, if you so wish.

18 Heating your new house

A new house should be designed to last for at least fifty years and in that time the popular fuels of the moment—oil and gas —will become extremely scarce and will quite probably not be available for domestic heating. The house should not therefore be tied to any one fuel, and should be designed with a flexible approach so that, if necessary, the type used can be changed over the years. I believe that the fuels which will come into prominence again are electricity and coal. The estimated coal deposits in this country are adequate for some three hundred years at our present consumption rates—longer if we reduce our consumption as proposed in this book.

Advanced heating systems such as heat pumps, described briefly in chapter 20, are gradually being developed and if you have a very enterprising architect he may like to suggest that you have one installed. Most houses, however, will have conventional heating systems and the remainder of this section deals only with these. We all have different likes and you may prefer the homely appearance and feel of a gas or solid-fuel fire or you may be happy with no obvious focal point in the room so long as it is warm and comfortable.

Small terraced or town house
This house will have such a low power consumption that, despite the previous remarks, the clear choice is to go all-electric. All-electric houses built in the sixties and seventies unfortunately obtained poor reputations and it was entirely because they had little or no insulation and consequent high power consumptions.

The first suggestion therefore would be to have a controlled–output, off-peak electric storage heater in the living-room with perhaps a 1kW wall-mounted, electric fire for very cold days. If the latter has a coloured bulb built in to it then a focal point and cosy glow will be obtained. The water would be heated by off-peak electricity in a large, insulated, storage tank. An oil-filled electric radiator or towel-rail, connected to an off-peak circuit for economy, can be fitted in the bathroom for drying towels. The bedrooms need no heating, because the heat from the ground floor will rise into them, and the kitchen should be warm enough with the heat coming from the refrigerator, cooker and other appliances.

There is little point in having expensive underfloor heating systems installed, partly because we have already decided to spend on insulation rather than on heating systems and partly because the off-peak tariff, which is now available to new consumers overnight only, makes these systems inefficient: they are at their warmest in the morning and coolest in the evening. As an alternative, you may prefer a high-efficiency gas fire in the living-room in place of the storage heater and electric fire.

Three-bedroomed semi-detached house

Like the smaller town house, this could also be all-electric with the main heating from a controlled-output storage heater and electric fire. An alternative would be to have a high-efficiency gas fire in the living-room. The dining area would need no heating in either case and, again, off-peak electricity could heat the water and a towel-rail. The kitchen and bedrooms need no separate heating. A further alternative would be to have a high-efficiency solid-fuel fire in the living-room. These fires burn in special enclosed grates and your architect will suggest suitable models.

Four-bedroomed detached house

It is usual to provide a chimney in this size of house and underfloor draught for the grate should be built in. The living-

136

room could then be heated by a controlled-output off-peak storage heater or a gas or solid-fuel fire. A smaller gas fire or storage heater would suffice in the dining-room or dining area. Off-peak electricity could supply the hot water and heat a towel-rail in the bathroom. Again, neither the kitchen nor bedrooms need separate heating.

Five-bedroomed detached house

A compact house would be heated very similarly to the four-bedroomed house but there is usually, in addition, an extra ground-floor room. If this is only used occasionally, say as a study, then a wall-mounted electric or gas fire would suffice. For more frequent use, as a children's playroom for example, an off-peak, electric, storage heater would be more suitable and safer.

A larger style of house would have a simple central heating system run from a small boiler, feeding skirting convectors in the ground-floor rooms and supplying the hot water. An alternative would be to have a high-efficiency gas, or solid-fuel fire in the main living-room, with a built-in back-boiler. The hot-water cylinder should be fitted with an immersion heater on an off-peak circuit for summer use. Again, neither the bedrooms nor kitchen should need to be heated from the system.

Bungalows

In a bungalow there is, of course, no heat rising from below to warm the bedrooms. The bedrooms will therefore require some form of heating. Small, off-peak storage heaters or electric fires would probably suffice. A larger bungalow would have the skirting convector system extended to include the bedrooms.

It is important to realise that a correctly insulated house loses heat relatively slowly and that complete reliance can be placed on individual heaters. The house will still be warm in the evening even if it has been unoccupied all day. Heaters can be switched on, if necessary, when you return, because the house will respond quickly to internal heating. This is sometimes difficult to appreciate if you are used to central heating in an un-insulated house, which has to be switched on some time before you return to give it time to warm up the cold house. For a large part of the year no heating will be needed in your insulated house.

19 Heating your present house

Having completed the insulation work it is sensible to make the best use of the heating system you have, as this avoids further immediate expense. In the long term it may be necessary to consider a change and these points are borne in mind with the following suggestions.

Gas fires
If you have gas fires in the downstairs rooms, then these are ideal. They can be switched on when you return to the house and will rapidly warm the house, even in winter. Some gas fires are more efficient than others—a check in the Consumers Association magazine *Which?* will tell you the best types to use.

Electric off-peak storage heaters and underfloor heating
If you have off-peak electric heating then you will find a remarkable improvement in the effect of the system. An insulated house allows this type of heating to work properly. For the first time you will find that the controls can be turned down, and if the system is automatic it will probably switch itself off for a large part of the time.

You should check that all the controls are working correctly —particularly the thermostats. If there is a charge control operated by an outside thermostat it will need resetting to give a lower charge.

Electric fires

If you rely on electric fires, for a smaller house you will find that they can be quite satisfactory. Again, they need only be switched on when the house or room is occupied. In the winter, one bar will suffice, where two or three were needed the previous year, and the fire will be off altogether for a larger part of the year. Whilst gas is at the moment cheaper than electricity, gas fires are expensive to buy and install and a change is very likely not worth your while.

Solid-fuel fires

If you have open, smokeless-fuel fires and the grates have underfloor draught, like the 'Baxi', then these are quite satisfactory because they do not draw too much air from the house. You may find it more convenient to purchase a small, electric fire for use until the really cold weather comes along. Even on-peak electricity is not too expensive for heating, provided that it is only used for short periods. When the weather deteriorates you can light the open fire. While the electric fire is in use, a fire-screen should be used as described on page 85 to prevent draughts blowing down or up the chimney.

Open grates which do not have underfloor draught are *not* satisfactory because they draw too much air from the house and air is very expensive to heat. In an insulated house, most of the fuel used for heating purposes goes toward heating the air necessary for ventilation. It would be advisable to install a high-efficiency gas fire in the main living-room for a start. If the house is large, then gas fires would be installed in other ground-floor rooms as well. In a smaller house, an electric fire or fan heater could be used in the other room until you determine whether or not the heating is adequate.

Central heating

If you have a central heating system, radiators or warm air, this is quite all right. If these are controlled by room thermostats, then they will turn the system off automatically for quite long periods. However, should the system only have a thermostat on the boiler, you will have to partially or completely close some of the radiator valves to prevent the rooms from becoming too hot. The boiler will automatically run for shorter periods and use less fuel. It may be possible to turn the boiler thermostat to a lower setting provided that the hot-water temperature is still adequate. In the summer it may be more economical to use an immersion heater, especially if it is connected to an off-peak circuit, rather than have the boiler just heating the hot water.

Running the boiler with the radiators partially shut down will cause the boiler to cycle on and off fairly frequently. Most gas and oil boilers can be adjusted slightly by altering the size of the flame, to balance the output of the boiler with the radiators. On newer gas boilers the adjustment is by a screw, and on older boilers by weights, both in the regulator valve. On oil-fired boilers the adjustment is done by altering the jet size. Ask your plumber or local heating engineer to make the necessary adjustment. This will result in the boiler running for longer periods at a lower setting and will save undue wear and tear on the valve gear and controls.

Again, it is important to realise that an insulated house loses heat slowly. The effectiveness of your present heating system will be dramatically increased and it will be capable of heating the house in a much shorter time than previously. Fires and heaters need not be switched on until you return after being out because the house will still be warm. The time-switch on your central heating system should be similarly reset so that it only operates while you are actually in the house.

Portable gas and paraffin heaters give off water vapour as a result of burning the fuel. In a draughty house this is all blown away but in an insulated house it can lead to condensation problems where none need occur. These heaters should not be used, nor indeed are they required, in a properly insulated house.

20 Heating systems of the future

Solar-heating systems

If you place your hand on a slate or tiled roof which has the sun shining on it, you will find that the roof is quite hot. Solar-heating panels, at their simplest, consist of a central heating radiator panel mounted on the roof, which has to be predominantly south-facing. When the sun shines it becomes hot, water is pumped through which, in turn, becomes warm and this is used to heat the domestic hot-water supply. The panel is enclosed and covered with glass to reduce heat losses and more sophisticated systems use copper or steel tubes, and sometimes oil, as the intermediate fluid.

What are the advantages? Apart from the small running cost of the pump and control gear, the system does obtain free heat from the sun and it will reduce the cost of heating your hot water. What are the snags? Well, in the British Isles, the sun only shines sufficiently strongly to heat the panel in the summer—and not always even then, so the total saving for the year will be about one quarter of the present cost of heating your hot-water supply.

In chapter 3 we saw that by far the greatest amount of heat used in an un-insulated house goes towards heating the house. The proportion used for other purposes, including heating the water, is a fairly small amount. So that, if you spend your money on a solar heating system, you are going to save one quarter of a small amount of your fuel bills which is not very much. You would be far better off initially spending your money on insulation, where you can save up to 80 per cent of a much larger amount.

Heat pumps

Imagine that you build a refrigerator into your house wall, with the radiator at the back of the refrigerator inside the house and the refrigerator door outside the house, then take the door off and switch the refrigerator on. It will run continuously, cooling the air outside and giving off the heat which it has extracted from the air, to the inside of your house, via the warm radiator. This is the principle of a heat pump. A correctly designed system is quite capable of extracting heat from the air outside even when it is freezing.

It can be used to extract heat from all sorts of intermediate sources. Suitably piped, it can extract heat from the outer leaf of your insulated walls, which in turn recover heat from the surrounding air. It can be used to extract heat from solar-heating panels on days when the sun is hardly shining. It converts a lot of low-temperature heat from outside the house into a small amount of high-temperature heat which can be usefully used inside the house.

This is the real heating system of the future but its introduction depends upon the house being adequately insulated in the first place.

Air-conditioning systems

Once your house has been correctly insulated, the greatest proportion of heat required to keep the house warm is used in heating the air necessary for ventilation. In a good air-conditioning system, the used warm air is passed over a radiator or pipe system where it heats the incoming fresh air. The amount of heat lost in the stale air going out of the house is then reduced to a minimum.

The heating system of the future will comprise a well insulated house, along with a heat pump and a heat recovery system.

The author wishes to acknowledge technical information received from the following companies:

DIY GLAZING SYSTEMS

D D Home Improvements, P.O. Box 4, Gerrards Cross, Buckinghamshire.
Davos Double Glazing, Washington, West Sussex.

SASH WINDOW FITTINGS

Royde and Tucker Ltd, 117 Stoke Newington Road, London N 16.

CAVITY WALL INSULATION

Rentokil Ltd, Felcourt, East Grinstead, West Sussex.
Megafoam Ltd, Morley Road, Tonbridge, Kent.
Eljay Insulation Ltd, Tannery House, Tannery Lane, Send, Woking, Surrey.
Cape Insulation Services, Rosanne House, Bridge Road, Welwyn Garden City, Hertfordshire.

EXPANDED POLYSTYRENE

Vencil Resil Co Ltd, Arndale House, Arndale Centre, Dartford, Kent.

PANELLING BOARDS

Cape Boards and Panels Ltd, Iver Lane, Uxbridge.